Microprobe Analysis as Applied to Cells and Tissues

Microprobe Analysis as Applied to Cells and Tissues

Proceedings of a Conference at Battelle Seattle Research Center
Seattle, Washington, U.S.A., April 30—May 2, 1973

Edited by

Theodore Hall

Cavendish Laboratory and
Biological Microprobe Laboratory
University of Cambridge, England

Patrick Echlin

Botany School and
Biological Microprobe Laboratory
University of Cambridge, England

Raimund Kaufmann

Institute of Physiology
Department of Clinical Physiology
University of Dusseldorf, Germany

1974

ACADEMIC PRESS · LONDON · NEW YORK

A Subsidiary of Harcourt Brace Jovanovich, Publishers

ACADEMIC PRESS INC. (LONDON) LTD.
24/28 Oval Road,
London NW1

United States Edition published by
ACADEMIC PRESS INC.
111 Fifth Avenue
New York, New York 10003

In accord with that part of the charge of its founder, Gordon Battelle, to assist in the further education of men, it is the commitment of Battelle to encourage the distribution of information. This is done in part by supporting conferences and meetings and by encouraging the publication of reports and proceedings. Towards that objective this publication, while protected by copyright from plagiarism or unfair infringement, is available for the making of single copies for scholarship and research or to promote the progress of science and the useful arts.

Library of Congress Catalog Card Number: 74 2147
ISBN: 0 12 319050 9

PRINTED IN GREAT BRITAIN BY
T. & A. Constable Ltd., Edinburgh

CONTENTS

SESSION I

Physical Bases, Techniques and Instrumentation of the Different Methods of Microprobe Analysis

Contents

SESSION II
Preparation Techniques

SESSION III
Quantitation and Special Techniques

Contents

SESSION IV

Applications of Microprobe
Analysis to Cells and Tissues

Contents

Dr. Rein Beeuwkes
Department of Physiology
Harvard Medical School
25 Shattuck Street
Boston, Mass.   02115

Professor Breck Byers
Department of Genetics
University of Washington
Seattle, Wash.   98195

Dr. James Coleman
Department of Radiation
Biology and Biophysics
University of Rochester
School of Medicine and
  Dentistry
Rochester, N. Y.   14642

Dr. Adolf Dörge
Physiologisches Institut
  der Universität München
8 München
Pettenkofer Strasse 12
Germany

Dr. Patrick Echlin
University of Cambridge
Botany School
Downing Street
Cambridge CB2 3EA
England

Dr. Wolfgang Fuchs
II.Physiologisches Insti.
  d.Univers.d.Saarlandes
665 Homburg/Saar
Paracelsusstr.12
Germany

Dr. Georges Fülgraff
Klinikum der Universität
  Frankfurt
Zentrum der Pharmakologie
Abteilung II
6 Frankfurt/Main 70
Theodor-Stern-Kai 7
Germany

Prof. ag. Pierre Galle
Faculté de Médecine de
  Creteil
Laboratoire de Biophysique
6, Rue du Général Sarrail
94.000, Creteil, France

Dr. Juergen Gullasch
Siemens A.G.
75 Karlsruhe
Siemens A.G., ABT.E63
Rheinbruckenstr.50
Germany

Dr. Theodore Hall
Cambridge University
Cavendish Laboratory
Free School Lane
Cambridge CB2 3RQ, England

Participants

Dr. Kurt F. J. Heinrich
Analytical Chemistry Div.
National Bureau of
 Standards
Washington, D.C.   20234

Dr. William M. Henry
Battelle
Columbus Laboratories
505 King Avenue
Columbus, Ohio   43201

Dr. Franz Hillenkamp
Gesellschaft fur Strahlen-
 und Umweltforschung
Abteilung fur Koharente
 Optik
8042 Neuherrberg
Ingolstadter Landstr. 1.
Germany

Prof. Thomas E. Hutchinson
Department of Chemical
 Engineering and Materials
 Science
151, Chemical Engin. Bldg.
University of Minnesota
Minneapolis, Minn.   55455

Dr. F. Duane Ingram
Department of Physiology
 and Biophysics
University of Iowa
Iowa City, Iowa   52240

Dr. Raimund Kaufmann
4000 Düsseldorf
Institute of Clinical
 Physiology
University of Düsseldorf
Moorenstr. 5, Germany

Mr. Rainer Koenig
Battelle-Institut e.V.
6000 Frankfurt/Main 90
Postschliessfach 900160
Germany

Dr. André Läuchli
Fachbereich Biologie
 - Botanik
Technische Hochschule
 Darmstadt
D-6100 Darmstadt
Germany

Dr. Claude Lechene
Department of Physiology
Harvard Medical School
25 Shattuck Street
Boston, Mass.   02115

Mr. Kenneth W. Marich
Research Associate
Stanford University
 Medical Center
Department of Pathology
Stanford, Calif.   94305

Dr. Richard S. Morgan
Department of Biophysics
511 Life Sciences Bldg.
Pennsylvania State U.
University Park, Penn.
 16802

Dr. Marjorie D. Muir
Department of Geology
Royal School of Mines
Imperial College of Science
 and Technology
Prince Consort Road
London SW7 2BP, England

Participants

Dr. Hans Neurath
Chairman, Department of
 Biochemistry
J405 Health Sciences SJ-70
University of Washington
Seattle, Wash. 98195

Dr. W. A. P. Nicholson
Institut für Medizinische
 Physik der Universität
 Münster
44 Münster/Westfalia
Hüfferstrasse 68
Germany

Dr. Poen S. Ong
Section of Experimental
 Pathology
Texas Medical Center
University of Texas M.D.
 Anderson Hospital and
 Tumor Institute
Houston, Texas   77025

Dr. Barbara Panessa
EDAX Laboratories
Division of EDAX Int. Inc.
4509 Creedmoor Road
Raleigh, N. Carolina 27612

Dr. John C. Russ, Director
EDAX Laboratories
Division of EDAX Int. Inc.
4509 Creedmoor Road
Raleigh, N. Carolina 27612

Dr. Gerhard Schimmel
Battelle-Institut e.V.
6000 Frankfurt/Main 90
Postschliessfach 900160
Germany

Dr. Arthur R. Spurr
Professor
College of Agriculture
University of California
 at Davis
Davis, California 95616

Dr. R. Taylor
Cambridge Scientific
 Instruments, Ltd.
Chesterton Road
Cambridge, CB4 3AW
England

Dr. Andreas Thaer
Battelle-Institut e.V.
6000 Frankfurt/Main 90
Postschliessfach 900160
Germany

Dr. Ronald R. Warner
Department of Physiology
Yale University
333 Cedar Street
New Haven, Conn.   06510

Dr. William R. Wiley
Battelle
Pacific Northwest Lab.
P. O. Box 999
Richland, Wash.   99352

Dr. Aurelio Zerial
Battelle
Geneva Research Centre
7, route de Drize
1227 Carouge-Geneva
Switzerland

Participants

Bottom Row (L-R)	Middle Row (L-R)	Top Row (L-R)
Dr. G. Schimmel	Dr. C. Lechene	Dr. F. D. Ingram
Dr. A. Thaer	Prof. A. P. Galle	Mrs. F. D. Ingram
Prof. T. E. Hutchinson	Dr. R. R. Warner	Dr. W. M. Henry
Dr. P. Echlin	Dr. P. S. Ong	Dr. W. A. Nicholson
Dr. J. C. Russ	Mr. K. W. Marich	Dr. M. D. Muir
Dr. R. Kaufmann	Dr. W. Fuchs	Dr. R. Beeuwkes
Dr. A. R. Spurr	Dr. R. Koenig	Dr. K. F. J. Heinrich
Dr. A. Zerial	Dr. H. Neurath	Dr. R. Taylor
	Dr. G. Fülgraff	Dr. R. P. Schneider
	Dr. J. Gullasch	Dr. W. R. Wiley
	Dr. A. Läuchli	Dr. F. Hillenkamp
		Dr. J. Coleman
		Dr. R. S. Morgan
		Dr. T. Hall
		Dr. A. Dörge

PREFACE

This book is an outcome of a conference on Microprobe Analysis as Applied to Cells and Tissues, held
April 30 through May 2, 1973, at the Battelle Seattle
Research Center, Seattle, Washington. In a field of
rapidly growing potential, the intention was to bring
together a small group of experienced workers to concentrate on the latest developments and on the current
difficulties. The conference was held without parallel sessions and with ample time for discussion. Since
many (though certainly not all) of the leading bio-
logically-oriented microprobe specialists were present,
the members of the conference felt that prompt publication of the proceedings would be useful.

A special effort has been made to present the
discussions which followed each paper as well as the
general discussion on the problems of preparation of
biological specimens for microprobe analysis. In some
ways these discussions bring out the current difficulties more strikingly than do the prepared papers. The
editors were therefore anxious to preserve this material, but were concerned not to delay the publication
on this account. The procedure finally adopted was
to send each author the tape recording of the discussion of his (or her) paper, and to have the author use
this material as he saw fit, leaving out what seemed
not worthwhile and choosing his own format and degree
of paraphrasing. We wish to make this clear especially with respect to those papers which do not conclude
with a section labelled "Discussion"; generally these
authors have chosen to incorporate the gist of the
discussion into the text of the paper itself. The
general discussion was prepared by the editors, also
from a tape recording. But it should be recognised
that most of the discussions have not been presented
Verbatim, nor has the rendition of each piece of

Preface

discussion been checked with its originator.

In preparing these proceedings the editors were assisted by Dr. A. Thaer, Battelle Institut e.V., Frankfurt, who together with Dr. R. Kaufmann was responsible for the organization of the conference.

The conference and the publication of the proceedings was made possible by the generous support provided by the Battelle Institute, Life Sciences and Engineering Sciences Programs. The organizers and editors would like to express their gratitude to Dr. W. R. Wiley and Dr. S. D. Ban, coordinators of the Life Sciences and Engineering Sciences programs sponsored by the Battelle Institute, and to Dr. H. Neurath, University of Washington for their active interest and support.

The help of the staff of the Battelle Seattle Research Center, in particular of Mr. L. M. Bonnefond and his staff in organizing the conference, and of Mr. A. W. Roecker and his staff in the editorial work and in preparing the camera-ready typescript for the delivery to the publisher, is also gratefully acknowledged. Finally we wish to thank the staff of Academic Press (London) for their contribution in bringing the book to publication.

December 1973   The Editors

Organizing Committee

Prof. R. Kaufmann   Dr. A. Thaer
Inst. Clinical Physiol   Battelle Institute e.V.
University of Dusseldorf   Frankfurt/Main
Germany   Germany

Conference Coordinator

L. M. Bonnefond, Battelle Seattle
Research Center, Seattle, Washington

INTRODUCTORY REMARKS
OF THE CONFERENCE CHAIRMAN

T. A. HALL

First I would like to express the thanks of all of us for the excellent facilities which have been made available to us here.

Among the conference documents which have been given to us, at various points, there are suggestions as to what might appropriately go into these introductory remarks. It is mentioned that it would be good to discuss the scope and purpose of the conference, to review the history of the application of microprobe analysis to biology, to consider the so-called "state of the art" of microprobe techniques, and to consider current problems and future developments, and this ground is to be covered in fifteen, or preferably in ten minutes.

Well, our purpose seems definable easily enough. Microprobe analysis is a means of determining the chemical composition of microvolumes, *in situ*, and our purpose here is to exchange all knowledge and ideas which will enable us to do this as effectively as possible in biology. As to the scope of the conference, the first point that appears from the programme is in the range of techniques that we are to consider. This is well beyond the specific technique of electron-probe, X-ray microanalysis. Most of us here, who are specialists in the electron-probe X-ray method, must recognise that there exist quite a few other microprobe techniques, including several which will be discussed at this conference, namely optical fluorescence and other optical analytical methods, X-ray fluorescence analysis, ion beam analysis, cathodoluminescence

analysis, and laser probe analysis. Here we specialists should listen and learn about the capabilities
of the other methods, so that in the future we can
give a sensible rather than a provincial judgement of
the best approach to a given biological problem. I
am sure that the conference will help the electron-
probe specialists, like myself, to look beyond the
elaborate instruments we are chained to; it will teach
us a great deal about what the other microprobe techniques have to offer.

A second point about scope emerges from the programme. Most of the electron-probe specialists here
are already fairly familiar on the whole with each
other's line of work. Therefore the papers must not
be lengthy expositions that will consume the entire
available time; on the contrary the papers should be
jumping-off points, leaving much time for discussion.
The discussion should be critical, not necessarily
severe but thorough and detailed. So the scope of the
conference is emphatically meant to include detailed,
specific evaluation and criticism of the various lines
of work which will be introduced by the speakers.

As to history, "state of the art" and future developments, I shall restrict my comments to electron-
probe X-ray microanalysis because I just don't know
enough about the other techniques to speak usefully
about them, and I shall compress past, present and
future into one passage for the sake of brevity.

The first functioning electron-probe X-ray micro-
analysers were announced at about the same time by
Castaing in France and Borovskii in the Soviet Union.
Castaing's design was taken up commercially by Cameca
around the year 1956. This was a static-probe instrument: scanning images were not available, and an integral optical microscope provided the only way of
seeing the specimen. The diameter of the focussed
probe was usually about 1 μm, or at best about ½ μm.
X-ray spectroscopy was performed by means of curved

diffracting crystals (in fact, Castaing had originally tried to use flat crystals, but the X-ray collecting power of these was hopelessly low). The limit of detectability for a given element was generally of the order of 10^{-14} g. Even with an instrument of this type certain biological studies were possible. But there were severe limitations due to poor spatial resolution of the X-ray analysis, poor visualisation and the difficulty of placing the probe in the desired place, and above all, low X-ray collecting power, which necessitated high beam currents which often destroyed or severely damaged biological specimens.

A major improvement, in the late 1950's, was the provision of beam scanning facilities and of scanning electron images in the microprobe analyser. For many specimens, essential detail could be seen in the scanning electron image better than with the integral optical microscope. And the static probe could now be positioned by reference to the scanning image formed on the persistent screen of a cathode-ray display tube. This was a much more positive method of localisation of the probe.

The next big step was the introduction, around 1960, of fully focussing, linear-track diffracting-crystal spectrometers with an X-ray collecting efficiency much higher than was previously available. This made it possible to work with lower beam currents and finer probes. The limit of detectability was brought down to something like 10^{-16} g of element in thin specimens, a most significant advance since one might have this amount of an element even at modest concentration within, say, an individual mitochondrion.

But it was not yet possible to take full advantage of the improved sensitivity because, at that time, there was no good way of seeing organelles like mitochondria in specimens within microprobe analysers. In order to avoid artifacts it was definitely preferable to analyse unstained tissue sections, and a viewing

method with very high contrast was needed. The transmission electron image provided the only generally
adequate viewing mode, and the requisite instrumental
advance came in the mid-1960's in the form of two alternative approaches. The "EMMA" scheme combined high-
quality X-ray spectroscopy with "conventional" electron microscopy. The alternative approach was to fit
high-quality X-ray facilities to scanning electron
microscopes and to make use of transmission scanning
images; one could then take advantage of the finer
probes available in SEM's in order to get images with
good spatial resolution, and in order to localise microprobe analyses to still smaller volumes.

The most recent in the series of major advances
has been the development of the "energy-dispersive"
Si(Li) X-ray spectrometers. These are very convenient in operation and they provide simultaneous analysis for all elements over a wide band of wave lengths.
Most significantly, their very high X-ray collecting
efficiency brings the limit of detectability down now
to something like 10^{-18} g of element in ultrathin
specimens.

As a result of all of these advances, we have
now been provided with the means to microanalyse thin
or ultrathin sections on the basis of specimen images
with spatial resolution in the range 1-10 nm (10-100 Å),
with spatial resolution of the X-ray analyses possibly
as good as 100 nm or slightly better in favourable
cases, and with a sensitivity that may detect 10^{-18} g.
Where do we go from here?

We know that we shall be getting brighter electron guns. Lanthanum hexaboride or similar guns, and
field emission guns, will be coming into routine use.
The brighter guns, and electron lenses with lower
spherical aberrations, will certainly open the way to
X-ray microanalysis with still finer spatial resolution, and to the measurement of even smaller elemental amounts. We can count on steady progress as well

in theories of quantitative analysis.

However, advances of a different kind will be needed to take care of the problem which is now felt to be the most pressing technical difficulty in microprobe biology. Most of us are not too unhappy with the spatial resolutions and sensitivities now available, and we do not lack methods of analysing quantitatively the specimen which is "seen" by the probe. The pressing problem is to ensure that this specimen, as it exists during the microanalysis, is sufficiently similar to the original material that we want to analyse. We have to control the changes occurring during specimen preparation, and we have to control beam damage and other environmental effects within the microprobe instrument. These problems will occupy a central place, and will be considered very extensively, at this conference.

Finally, one more remark about the scope of this meeting; the instrumentalists among us must keep our feet on the ground. We should not try to institute developments just because they seem nice to us, without reference to the actual needs of biologists. This is basically an instrumental conference (it must be since its title bears the name of a class of instrumental techniques), but an _essential_ part is to be played by the reports we shall hear from the applications side, the reports of real biological research. These reports should help to keep our directions of development sensible.

That is all I had in mind to say. Shall we begin?

RECENT RESULTS IN THE DEVELOPMENT
OF A LASER MICROPROBE [1]

F. HILLENKAMP*, R. KAUFMANN**, R. NITSCHE*,
E. REMY*** AND E. UNSÖLD*

*Gesellschaft für Strahlen- u. Umweltforschung mbH,
Abt. f. Kohärente Optik, 8042 Neuherberg, Ingol-
städter Landstr. 1.
**Institut für Klinische Physiologie der Universität
Düsseldorf, 4 Düsseldorf, Moorenstrasse 5.
***Firma Biotechnik, 8 München 40, Georgenstrasse 22.

INTRODUCTION

Several methods of sampling cells and tissue for
the purpose of a microprobe analysis have been suggested and used. Focussed laser beam is one of them.
The spatial resolution of this sampling technique,
though in principle limited only by optical diffraction to about 0.5 µ, is actually determined by the
interaction process between light and the sample, and
the sensitivity of the analytical method used.

This contribution concentrates essentially on the
mass-spectrometric analysis of laser-sampled objects.
This method is still in a rather early stage of development and only limited information can be given on
the analysis of biological samples as yet. The results so far achieved on the sampling of cells and
tissue sections and on the mass-spectrometric analysis
of metallic layers evaporated onto glass substrates,
indicate that the method will be applicable to biological objects, particularly, for the detection of the
inorganic electrolytes in the cell.

---

[1] Sponsored by the Stiftung Volkswagenwerk

GENERAL OUTLINE OF A LASER MICROPROBE

Laser-Optical Instrumentation

The laser-optical part of the set-up must serve two purposes: focussing of the laser beam to a diffraction limited spot and imaging the sample with maximum resolution and magnification in order to get an unambigous correlation between the sampled area and the different cell structures. Among the various arrangements possible, we have chosen to use the same optical system for observation and focussing as shown in Figure 1. The sample is located at the lower side of a cover slide with no microscope slide underneath (Figure 2). This arrangement allows for a maximum

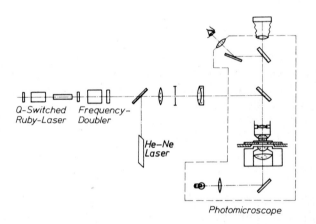

Q-Switched Frequency-
Ruby-Laser Doubler

He–Ne
Laser

Photomicroscope

Figure 1. *Schematic diagram of the laser-microscope.*

resolution and minimum focus through the use of oil-immersion objectives as well as observation in light- and dark-field illumination, phase-contrast and interference contrast, if suitable optical condensers are used. The cover slide simultaneously acts as a vacuum window to separate the optical part from the vacuum containment of the mass-spectrometer. If limited resolution suffices, top illumination is used and for

cases of maximum resolution, the optical condenser below the sample must be shifted and replaced by an ion-
optic for every shot.

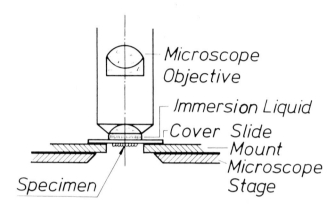

Figure 2. *Specimen support.*

Principles of Analysis

Light-Emission-Spectroscopy is one of the possible methods for analysing the sampled material. Several groups have developed this method during the
last decade. It's promises and limitation will be
discussed by Dr. Marich in the next paper. The sensitivity of the emission spectroscopy appears to be
limited to sampled objects of about the size of a
whole cell, and we have therefore started to look into
mass-spectrometry as an alternative method because of
its inherently higher sensitivity. Quadrupole-mass-
filters are suited for this purpose because they can
be trimmed to a high transmission leading to high absolute senstitivity. On the long run, TOF-spectrom-
eters appear to be more attractive as they allow the
registration of a whole mass spectrum for each single
shot. The mass-spectrometric analysis of a laser
generated plasma will primarily allow for the detection of the various isotopes, while the chances for
the measurement of organic molecules or fractions
thereof seem to be rather low. Laser-induced

microfluorimetry appears more suited for the analysis
of organic molecules. The availability of tunable
UV-laser sources in the wavelength region between 200
and 300 nm with powers and pulse times suitable for
time resolved microfluorimetry should start new activities in this field. Spatial resolutions of 0.2 μ
should be easily achievable and the powers available
should make it possible to detect even fluorescence of
low quantum yield, provided it can be separated from
the background.

RESULTS

*Light-Sample Interaction, Optimal Choice of Laser
Parameters*

As a first series of experiments, we investigated
the interaction between laser-light and cells for the
special case of microirradiation. The detailed results
of these experiments have been published elsewhere
(Hillenkamp *et al.*, 1973; Kaufmann *et al.*, 1972) and
only a short summary of the most important aspects of
them is given here:
a) Suitable optics and a proper choice of
wavelength and pulse time make it possible to repro-
ducibly achieve sampled areas of 1 μ diameter or less.
Under optimum conditions the exact magnitude of the
laser power is not a critical parameter (Figures 3, 4,
and 5).
b) The lack of absorption in the visible and
near UV of almost all cell structures- hemoglobin and
melanin granules being among the rare exceptions -
does not allow for a simple thermal process as the
primary interaction process. Laser powers in the focus must be high enough for the occurrence of non-linear processes such as two photon absorption.
c) Because of b) light in the visible is not
suited to sample the vast majority of cells and tissue.
A sharp "explosion" threshold is the cause for very
unreproducible results (Figure 6). Laser light in the

4

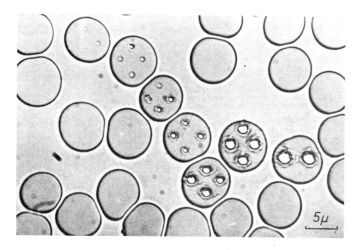

Figure 3. *Laser-perforated red blood cells; energy doubled from cell to cell* λ = 694 nm; T = 20 ns.

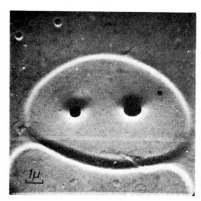 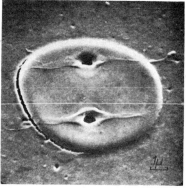

Figure 4. *Scanning-electron-microscope picture of perforated red blood cell* λ = 694 nm; T = 1 ms

Figure 5. *Scanning-electron-microscope picture of perforated red blood cell* λ = 694 nm; T = 20 ns

near UV at 347 nm was used which resulted in a very reproducible sampling of spatial resolution of 1 μ or better for a large variety of cells and tissues tested (Figure 7). The exact mechanism of the interaction is not yet fully understood.

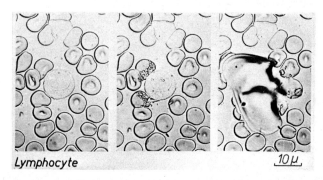

Figure 6. *Explosion of lymphocyte after irradiation at λ = 694 nm; T = 1 ms.*

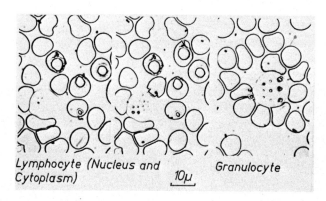

Figure 7. *Laser-perforated white blood cells λ = 347 nm; T = 20 ns.*

As a result of the experiments mentioned above we decided to continue our work with a frequency doubled ruby-laser at a wavelength of 347 nm. Because of the frequency doubling process Q-switched pulses of 20 ns duration had to be used, though the use of somewhat longer pulses seems to be preferable in order to avoid the mechanical effect of shock waves. The expected further development of tunable lasers with pulse times of about 1 μs in this wavelength region will however most probably lead to a better choice of the laser.

Development of a Laser Microprobe

Investigation of the Laser-Generated Plasma

The sensitivity of mass spectrometers and especially their resolution depend critically on the parameters of the laser produced plasma, particularly on the energy distribution of the ions. The results obtained from measurements of the integrated ion current from laser-produced microplasmas of metallic and organic samples indicated that the average ion energy was below 100 eV.

Ion energies of less than 100 eV are well suited for an analysis in quadrupole-mass-filters and we therefore went on to attach such an instrument to the laser microscope (Figure 8). The attachment of the

Figure 8. *Microprobe, consisting of laser-microscope and quadrupole-mass-filter.*

rigid, bulky and vibrating vacuum system to the sensitive optical instrument took some time because of the engineering problems involved. Special attention was given to the sample stage (Figure 9). The cover slide with the sample on its lower side is put onto a vacuum flange movable in the x-y plane. The microscope objective is movable in the z-direction for the

7

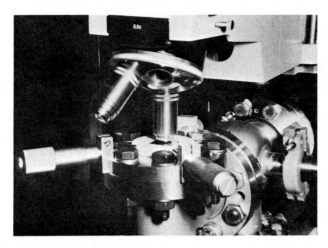

Figure 9. *Sample stage of microprobe.*

purpose of focussing. Aluminum films of 0.4 μ thickness, evaporated onto the cover slides were used as samples in the experiments described in the following. The sampled area was 2.5 - 3 μ in diameter in all cases (perforations of 1 μ in diameter were easily achieved with a x100 objective, but a x40 objective without oil immersion was used in these experiments for the ease of experimentation).

Figure 10 shows a schematic diagram of the standard quadropole mass filter (Balzers AG.). Ions from the sample are transmitted through the condenser into the entrance aperture of the filter. Ions of suitable mass are transmitted and deflected through 90° onto the first dynode of an electron multiplier. The 90° deflection was introduced in order to keep the laser light from reaching the multiplier tube. However, the energy dispersion of the deflection capacitor interfered somewhat with the interpretation of the results.

A mass spectrum of the aluminum coated glass samples is shown in Figure 11. Besides aluminum, signals from all elements, contained in crownglass were detected. This fact - surprising at first - is easily explained by the scanning-electron-microscope pictures

Development of a Laser Microprobe

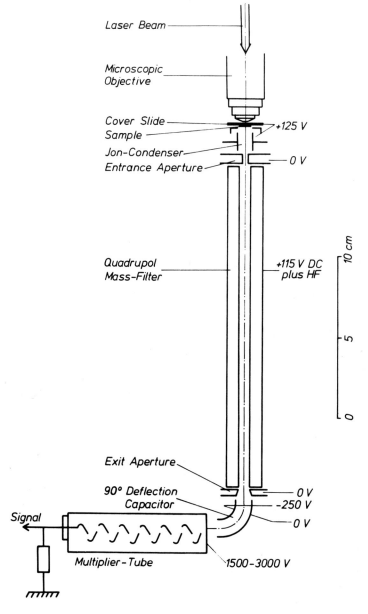

Figure 10. *Schematic diagram of ion source and quadrupole-mass-filter.*

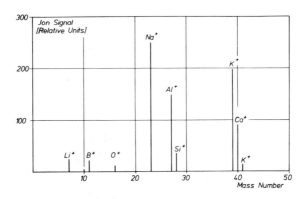

Figure 11. *Mass spectrum of aluminum on a glass substrate.*

Figure 12. *Scanning-elec-ton-microscope picture of laser perforations in an aluminum layer on a glass substrate.*

of the sampled areas as shown in Figure 12. A small amount of glass appears to be removed from the surface in addition to the evaporated aluminum. Smooth craters in the glass were found in all of the large number of investigated samples. The relative magnitude of the Na and K signals as compared to the Al signal is a consequence of their low ionization potential, a common fact in mass spectrometry. Their strength on the other hand is quite favourable for the microanalysis of biological samples. Quartz substrates will be necessary for a successful analysis in the future. They exhibited signals of alkaline ions smaller by

several orders of magnitude as compared to those from
glass substrates. Multiply charged ions were not
found. Their fraction is certainly less than 1%. A
rough estimate of the resolution so far achieved can
be gained from Figure 13. The silicon signal at mass
number 28 is just resolved in the wing of the aluminum signal which centers at the mass number 27. The
resolution of the instrument can - no doubt - be considerably improved by setting the filter parameters
to their optimum values. These values in turn depend
first of all on the energy of the filtered ions.

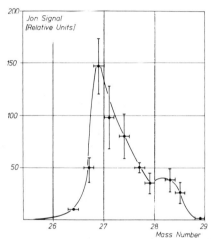

Figure 13. *Mass spectrum of aluminum and silicon.*

In order to measure the energy distribution of
the various ions we pulsed the potential of the sample
relative to the entrance aperture with a very steep
rising pulse ($\tau \sim 2$ ns) from +125 volts to 500 volts
at a variable time delay relative to the laser pulse.
The signal amplitude of Al-, Na- and K-ions as a
function of this delay time are plotted in Figure 14.
An analysis of this velocity distribution curves reveals surprisingly low average energies and energy
spreads of the ions produced (see insert of Figure 14).
This result - favourable for our purpose - is particularly surprising when compared to values of other
authors; values up to several keV have been reported

11

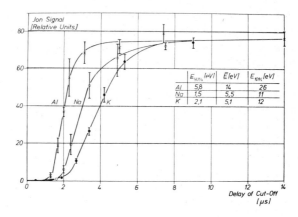

Figure 14. *Determination of ion energy by time-of-flight measurements.*

for laser generated metal plasmas. In all these cases, however, energy densities, several orders of magnitude higher than ours were used to irradiate large areas of solid metal surfaces. In our experiments, the ion energy of aluminum exceeds that of sodium and potassium by about a factor of 3. Different processes must therefore be responsible for the ion generation. The most plausible explanation is, that the aluminum plasma is formed through a direct absorption of the laser light, whereas the alkaline ions are emmitted from the glass after it had been melted and vaporized through heat conduction from the metal plasma. Figure 12, a scanning-electron-microscope picture of the irradiated substrate, taken at an angle of 60° to the normal, supports this theory. It clearly shows a dam of resolidified glass around the crater, indicating a melting-away rather than a burst-off process for the glass removal. One of the most important parameters is the absolute sensitivity of the method. The magnitude of the aluminum ion current, as measured at the tapped first dynode of the multiplier tube, indicated that about 1 in every 2500 aluminum atoms of the sampled volume was detected.

Assuming, that the ions are emitted isotropically into the half solid angle and taking the geometry of

the ion collecting system into account 3% of the atoms
of the sampled volume which could possibly enter the
filter therefore appear as detected ions. With the
ion energy as low as it turned out to be, this fraction can certainly still be improved by up to an order
of magnitude through the proper choice of the ion optics, imaging the ion source into the entrance aperture.

The sensitivity of the method for the detection
of Na and K is certainly higher than that for Al as
can be seen from the spectrum shown; the exact value
is hard to determine though, because we do not know
exactly how much sampled material we started with.

If we take the aluminum values for a conservative
estimate of the detection sensitivity of the method,
it turns out that about 10^4 atoms or $\sim 10^{-18}$g of the
analysed element are required in the sampled volume
in order to get a reasonable signal / noise ratio.
The relative sensitivity for a sampled volume of $0.2\mu^3$
is then about 1 ppm, very high indeed for an analytical method of a spatial resolution of 0.5 μ.

This high sensitivity is the most attractive but
not the only promise the method holds. With the resolution already achieved, different isotopes of elements can be detected thus facilitating the study of
the kinetics of exchange processes and the like.
Moreover the apparatus can easily be complemented by
a photon detector for a fluorescence assay of the organic molecules of the cell. So, as has been pointed
out previously, fluorescence techniques hold quite a
promise for future microprobe developments. The commercial availability of tunable laser sources down to
a wavelength of at least 250 nm, expected for the near
future will be an important step towards the realiza-
tion of such a technique.

The next series of experiments will concentrate
on organic samples. A few preliminary trials with

blood cells on quartz substrates have already resulted in well resolvable signals of Na and K. The question of whether all the molecules of the sampled volume will be broken down to single atoms as theoretical estimates predict or whether ionized fractions of
molecules lead to a background signal limiting sensitivity and resolution of the method will be of prime
interest in these experiments.

The development of a suitable time-of-flight
spectrometer will parallel these investigations.

REFERENCES

Hillenkamp, F., Kaufmann, R., and Remy, E., 1973,
*Proceedings of the Int. Symposium on Modern
Technologies in Physiological Sciences*, Academic
Press, New York.

Kaufmann, R., Hillenkamp F., and Remy, E., 1972,
Microscopica Acta , <u>73</u>, 1-18.

LASER MICROPROBE EMISSION SPECTROSCOPY
BIOMEDICAL APPLICATIONS

KENNETH W. MARICH

*Division of Histochemistry, Department of Pathology,
Stanford University Medical School, Stanford,
California. Present address: Information Science
and Engineering Division, Stanford Research Institute
Menlo Park, California.*

INTRODUCTION

With the development of sensitive analytical instrumentation (i.e. ion, electron and laser microprobes, atomic absorption spectroscopy, and fluorescence techniques) much interest has been generated in
the clinical significance of trace metals and metal-
bound compounds in the human body. Many biological
processes such as erythropoiesis, bone calcification,
and enzymatic reactions depend on specific metals.
Abnormal amounts of trace metals have been associated
with chronic diseases such as Hemochromatosis, Atherosclerosis, Rheumatoid Arthritis, Cystic Fibrosis and
Wilson's Disease to name a few.

Emission spectroscopic techniques have long been
employed for detection and quantitation of metals in
macro-biological samples (Stitch, 1957; Tipton *et al.*,
1963). Trace metal analysis of biological materials
requires selective microsampling and increased analytical sensitivity for determinations on microscopic
structures. Improvements in laser microprobe instrumentation have done much to satisfy these requirements.

15

MATERIALS AND METHODS

A block diagram of the laser microprobe apparatus is shown in Figure 1. The Q-spoiled ruby laser delivers a maximum energy of ~ 150 mJ in ~ 30 nsec. The

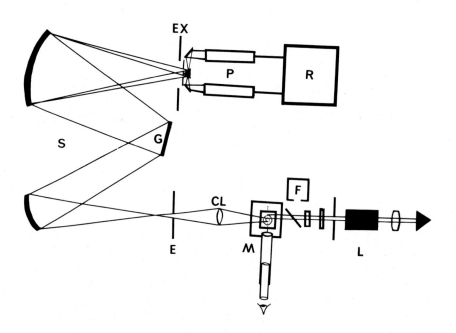

Figure 1. *Diagram of laser microprobe apparatus: laser cavity* (L), *filtered air cooling* (F), *microscope* (M), *light-collecting lens* (CL), *spectrograph* (S), *entrance slit* (E), *diffraction grating* (G), *exit slit* (EX), *photomultiplier tubes* (P), *and recording apparatus* (R).

principle employed is to vaporize a preselected portion of the sample with a laser beam focused by a microscope. Light from the plasma is collected and transmitted into an emission spectrograph and the intensities of specific spectral lines of elements in the vapor are recorded either photographically or photoelectrically.

Laser Microprobe Applications

Our original laser microprobe apparatus incorporating a two channel photoelectric detection system was designed and constructed as a collaborative effort between our laser group and engineers at Stanford Research Institute (Peppers *et al.*, 1968). Since then an improved instrument with a six-channel polychromator has been developed and is being used simultaneously with the original instrument for experimentation.

RESULTS AND DISCUSSION

In order to effectively use the laser microprobe for analytical purposes studies were made concerning matrix and atmospheric effects on optical emission (Marich *et al.*, 1970; Treytl *et al.*, 1971).

The matrix effects in laser-induced plasmas have distinct features, but in common with other forms of excitation suppression of optical signal by matrix material occurs as seen in Table I. When analyzing

TABLE I

Mean Percent Suppression[a] by Matrix of
Spectral Emission of Silver[b]
(Reprinted from Anal. Chem. 42, 1775 (1970)).

Matrix	Matrix Concentration (g/100ml)						
	0.01	0.06	0.28	1.4	7.0	28	35
Albumin	−4±17[b]	−15±19	32±13	80±4	88±2	—	—
Sucrose	0±15	17±14	34±12	58±7	81±4	—	—
Sodium acetate	−3±17	−6±17	37±9	37±12	68±5	88±2	90±2
Sodium sulfate	5±11	33±11	26±11	68±7	83±4	74±6	78±9

a. Mean of 10 samples ± std. dev.
b. Dried 40 nl samples (approx 400μ dia) containing 40 ng Ag, with and without matrix.

serum the protein matrix effect can be reduced by sample dilution, however the elements of interest are also diluted. In spite of the signal suppression by the protein matrix our best results with serum analysis were obtained without previous treatment or dilution. Analysis of human serum for calcium and magnesium is shown in Table II.

TABLE II

Determinations of 12 nl Samples of Blood Serum
(Means of 10 Analyses ± S.E.)

	Ca (mg%)		Mg (mg%)	
Patient	Atomic Ab.	Laser	Atomic Ab.	Laser
A	6.9 ± 0.03	7.2 ± 0.6	1.6 ± 0.01	1.6 ± 0.1
B	8.5 ± 0.03	8.8 ± 0.6	2.0 ± 0.02	2.1 ± 0.4
C	10.2 ± 0.06	10.1 ± 0.7	3.1 ± 0.02	3.0 ± 0.7

The effects of various atmospheres on signal intensity from laser-induced plasmas is shown in Table III. The results suggest that it may be advantageous

TABLE III

Effect of Atmosphere on Laser-Induced[a] Optical
Emission from Magnesium in Serum and Liver
(Integrated Photoelectric Data)
(Reprinted from Anal. Chem. 43, 1952 (1971)).

	Signal		S/B[d]	
Atmosphere	Serum[b]	Liver[c]	Serum	Liver
Vacuum	4.5 ± 1.3	1.6 ± 0.3	2.3 ± 1.0	3.7 ± 1.9
Air	20.3 ± 2.8	10.5 ± 1.6	1.8 ± 0.5	3.7 ± 1.0
Argon	34.1 ± 7.6	36.2 ± 7.9	1.1 ± 0.4	4.2 ± 1.1

a. Laser energy = 3.6 ± 0. 2mJ.
b. Human blood serum, air dried 40-nl samples, (approx. 400μ dia) mean for 10 samples ± std. dev.
c. Formalin-fixed, paraffin-embedded, and deparaffinized tissue sections, 5μ thick.
d. Signal-to-background ratio ± std. dev.

in certain cases to select an appropriate atmosphere
in order to optimize signal/background ratio (S/B) or
signal intensity. In general, however, the atmosphere
composition did not appear to effect the results sufficiently to warrant changing to atmospheres other than
air.

As an analytical instrument the laser microprobe
presents special noise problems. The microprobe
vaporizes a sample usually in the pg-µg range depending on the laser energy employed. While the duration
of the Q-spoiled laser beam is < 50 nsec and most of
the optical noise occurs during this interval, the
optically emitting plasma may persist for many micro-
seconds as can be seen in Figure 2. Results of our

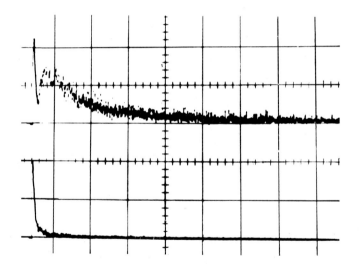

Figure 2. *Typical photomultiplier anode pulses*
Upper trace is signal channel at 3020 A° and lower
trace is background channel at 3015 Å. Horizontal
scale 1 µsec/div, vertical scale 4 mA/div. Conditions:
iron in NBS steel sample, air atmosphere, 4.0 mJ laser
energy. (Reprinted from Anal. Chem. 43, 1452 (1971)).

studies (Treytl *et al.*, 1971) show that significant reduction of optical noise can be accomplished with little loss of useful signal to give a marked increase of S/B ratio by time differentiating the emission to reject the early continuum as seen in Figure 3.

The time differentiated approach was used to determine detection limits of various metals in a

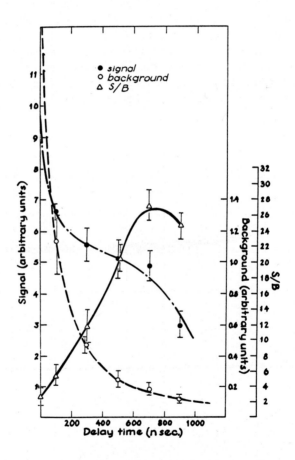

Figure 3. *Time differentiated behavior of optical emission from magnesium in aluminum foil at 2.5 mJ laser energy. (Reprinted from Appl. Spectroscopy 25, 376 (1971)).*

biological matrix (Treytl *et al.*, 1972) as presented in Table IV. As can be seen there is considerable variation for delay and integration times depending

TABLE IV

Optical Time Parameters and Detection Limits
of Metals in Organic Matrices

Element	Wavelength (Å)	Delay time (μ sec)	Intergration time (μ sec)	Detection limit (gram)
Li	6104	4	5	2×10^{-13}
Mg	2796	4	7	2×10^{-15}
Ca	3934	5	15	1×10^{-14}
Fe	3020	10	6	3×10^{-13}
Cu	3248	15	3	2×10^{-15}
Zn	2139	5	2	5×10^{-14}
Hg	2537	7	3	3×10^{-13}
Pb	4058	16	5	1×10^{-13}

Dried droplets of salts dissolved in 7% albumin used as samples for Mg and Ca, all other in 5% gelatin matrix. Laser energy 12-16 mJ; 25 nsec pulse duration. (reprinted from Anal. Chem. 44, 1903 (1972)).

on the element. Therefore when doing multi-elemental analyses, maximal sensitivity for *each* element can be achieved simultaneously if each channel is equipped with an independent time differentiated photoelectric detector. A block diagram of the photoelectric detection system is shown in Figure 4.

These detection limit values are compared with other sensitive micro-analytical methods in Table V. Where ultimate concentration sensitivity (ppm) is required and sample volumes may be rather large, atomic absorption, neutron activation, flame, arc, and spark emission methods or even wet chemistry techniques might be preferred. In applications where extremely small sampling resolution is required (< 5 μ dia)

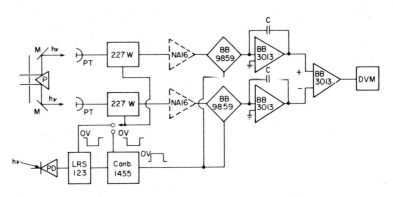

Figure 4. *Block diagram of photoelectric time
differentiating system.   P — 60° front surface reflecting prism; M — front surface reflecting plane
mirror;* PT — *photomultiplier tube (XP 1118, Amperex);*
227W — *gated integrator (227W, Lecroy Research Systems);* PD — *photodiode (4201, Hewlett-Packard);* LRS
123 — *discriminator (LRS 123, Lecroy Research Systems);*
Canb. 1455 — *variable delay logic pulse shaper (1455,
Canberra Ind.);* NA 16 — *linear amplifier (optional)
(NA 16, Hammer Elect. Co.);* BB 9859 — *linear gate
(9859, Burr Brown);* BB 3013 — *differential amplifiers
(3013, Burr Brown);* C — *0.01* μF *capacitor;* DVM — *digital voltmeter (111, Simpson).   (Reprinted from Appl.
Spectroscopy 25, 376 (1971)).*

electron or ion microprobe might be used. In other
applications where sampling size requirements are
more intermediate (5-150 μ dia) or where sample preparation is to be kept to a minimum, the laser microprobe seems to offer a desirable combination of sensitivity and spatial specifity.

Sampling resolution has been demonstrated by vaporizing heads of rat sperm as well as single blood cells.
Laser microprobe analysis of iron in one, two and
three red blood cells from four different patients is
shown in Figure 5. Detection and quantitation of iron
in single red blood cells ($\sim 10^{-13}$g) may provide valuable information related to hemolytic diseases and

possible cellular differentiation based on mean cell hemoglobin content.

TABLE V

Comparison of Capabilities of Highly
Sensitive Techniques for Elemental Analysis

Technique	Sample size	Detection Limits	
		ppm	gram
Ion microprobe	$10^{-11}g$ 1-5μ dia.	$10^0 \rightarrow 10^{-2}$	$10^{-17} - 10^{-19}$
Electron microprobe	$10^{-12}g$ 0.2-1μ dia.	$10^3 - 10^1$	$10^{-15} \rightarrow 10^{-17}$
Laser microprobe	$10^{-8}g$ 10-200μ dia.	$10^2 \rightarrow 10^{-1}$	$10^{-12} - 10^{-15}$
Neutron activation	mg-g	$10^{-3}-10^{-6}$	$10^{-10}-10^{-13}$
Atomic absorption	mg	$10^{-1}-10^{-5}$	$10^{-9} - 10^{-12}$

(Reprinted from Anal. Chem. 44, 1903 (1972)).

More recent instrumental improvements have re-sulted in lower detection limits for iron (0.6×10^{-13} grams). This increased sensitivity allows differen-tiation between red blood cells whose mean cell hemo-globins (MCH) differ by as little as 20×10^{-12} grams. Quantitative analysis of smaller differences in MCH may be possible with further technological improve-ments.

The continued use of gold for the treatment of rheumatoid arthritis as well as an agent producing obesity in mice prompted an investigation as to the effects of gold thioglucose on mouse fibroblasts *in vitro* (Herman *et al.*, 1972). The laser microprobe was used to selectively microsample and quantitate the gold uptake by the fibroblasts as a function of inoculation concentration and time as shown in Table VI

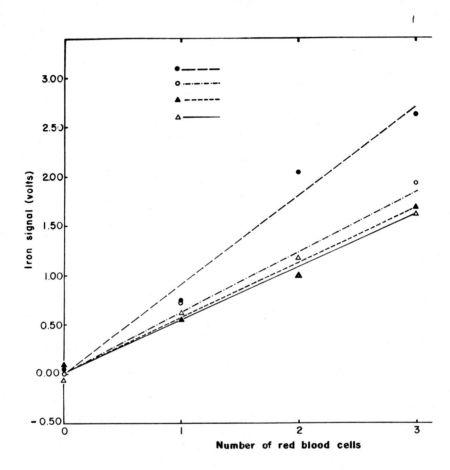

Figure 5. *Laser microprobe analysis of iron in single red blood cells from four different patients, (Mean of 10 samples).*

CONCLUSION

In conclusion I feel that the diverse applicability of the laser microprobe for microanalytical purposes in medical research and clinical analysis has been demonstrated. The advantages of the laser microprobe technique, including small sample size, little or no sample preparation, selective micro-sampling, and quantitative elemental analysis indicates the great potential of the instrument as an analytical tool with extensive biomedical applications.

I would like to acknowledge the fact that the work I have presented was the result of a collaborative effort with Drs. David Glick and William Treytl and was supported by Research Grant GM 16181 (to D.G.) from the National Institutes of Health, USPHS.

TABLE VI

Laser[a] Microprobe Analysis of Mouse Strain L Fibroblasts for Cellular Gold[b] Compared to Observed Cytopathic Effects[c] at 24 and 48 hours Incubation

Initial gold inoculation (mg Au/ml)	Incubation time			
	24 hours		48 hours	
	Gold (pg)/cell	CPE	Gold (pg)/cell	CPE
0.25	-1.4 ± 0.5	—	1.2 ± 0.9	—
0.50	0.0 ± 0.6	—	1.4 ± 0.9	—
1.00	-0.3 ± 0.7	—	1.9 ± 0.8	—
2.00	2.5 ± 1.0	—	7.3 ± 1.2	2+
4.00	4.5 ± 0.8	2+	11.3 ± 1.6	4+

a Laser energy = 13.9 ± 1.5 mJ
b Mean cellular gold (10 samples) - mean background gold (10 samples) ± std. dev.
c Cytopathic effect (CPE): - = < 30% round cells, 2+ = 50% round cells, 4+ ≅ 100% round cells.
(Reprinted from Exp. Mol. Path. 16, 186 (1972)).

REFERENCES

Herman, M. M., Bensch, K. G., Marich, K. W., and Glick, D., 1972, *Exp. Mol. Path.*, 16, 186-199.

K. W. Marich

Marich, K. W., Carr, P. W., Treytl, W. J., and Glick
D., 1970, *Anal. Chem.*, 42, 1775-1779.

Peppers, N. A., Scribner, E. J., Alterton, L. E.,
Honey, R. C., Beatrice, E. S., Harding-Barlow, I.,
Rosan, R. C., and Glick, D., 1968, *Anal. Chem.*,
40, 1178-1182.

Stitch, S. R., 1957, *Biochem.*, 67, 97-103.

Tipton, I. H., Cook, M. J., Steiner, R. L., Boye, C. A.
Perry, H. M., and Schroeder, H. A., 1963, *Health
Physics*, 9, 89-101.

Treytl, W. J., Marich, K. W., Orenberg, J. B.,
Carr, P. W., Miller, D. C., and Glick, D., 1971,
Anal. Chem., 43, 1452-1456.

Treytl, W. J., Orenberg, J. B., Marich, K. W., and
Glick, D., 1971, *Appl. Spectros.*, 25, 376-378.

Treytl, W. J., Orenberg, J. B., Marich, K. W.,
Saffir, A. J., and Glick, D., 1972, *Anal. Chem.*,
44, 1903-1904.

DISCUSSION

RUSS: It seems to me that emission spectroscopy could be done much more advantageously with something like an optical multichannel analyzer, the Vidicon.

MARICH: I think this technique might be a very good approach. The SSR Company in Southern California has developed what they call an OMA which is a silicon vidicon target having 500 channels. The limitations of this instrument are (1) wavelength sensitivity — its response only goes down to about 3500 A, while the majority of the most intense emission lines are below this wavelength, and (2) you are physically limited by the exit port of your instrument and its dispersion. For example, say that you wanted to evaluate a 1000 A bandwidth. Since there are 500 fixed channels, your resolution can be no better than 2 A per channel which means you could be inputing more than one line per channel. However, you could use it where you are only looking at a 100 A bandwidth and in that case you would have very good resolution (0.2 A per channel).

HEINRICH: I have the impression that we will find that one of the main problems in this whole micro-analysis is to stop the living cell and to keep things where they are and analyze them, and I would think that in that sense the laserprobe would have the unique possibility of analyzing a non-dehydrated cell and I wonder if this direction should not be empha-sized because there we have difficulty with instru-ments that require vacuum.

MARICH: We have done a limited amount of experimenta-tion on wet tissue. The problem with the generation of a plasma from wet tissue is the production of steam which makes it hard to differentiate the discrete spectral lines.

FÜLGRAFF: I want to ask Mr. Marich whether his system is better than atomic absorption? The figures you have given are for routine atomic absorption and not flameless atomic absorption. I believe in the case of the flameless technique the detection limits are about three orders of magnitude better and therefore this means that your system is equivalent to flameless atomic absorption.

MARICH: The table that I presented was in a publica-tion on the detection limits of our laser microprobe which came out in September, 1972. Since then there has been a lot of work done with the carbon rod atom-izer and atomic absorption. However, I have heard that there are many analytical problems with the carbon rod atomization technique but I do agree that the detection limits are getting better with atomic absorption spectroscopy.

LÄUCHLI: Is it possible with the laser microprobe to
detect elements such as carbon or nitrogen and if so
what is the sensitivity of them?

MARICH: We have not done studies on the sensitivity
of carbon but we are looking at carbon with the possible idea of using this element as a form of internal
standardization because it is native to all tissue.
We are presently trying to correlate the carbon optical emission with the amount of sample vaporized.   We
are also looking at nitrogen-hydride bands for the
same reason.

THAER: You mentioned the possible use of lifetime
measurements using Laser — excited fluorescence spectroscopy and that it might be possible to go down to
10 ns pulse width.   But even in this case, I think,
you would have to use convolution and deconvolution
techniques because you would have an overlapping of
the excitation peak with the emission peak.   Until
now nobody succeeded in using the well known lifetime
measurements in the application of this technique to
single cells and it might perhaps be more promising
to measure fluorescence quantum yield, fluorescence
intensity, fluorescence spectra and fluorescence
polarization.

HILLENKAMP: Yes, I think so, though this is not always true.   We have done Laser-excited fluorescence
studies on glasses for dosimetry purposes which is
something rather different.   But we could very easily
measure time decays and analyze them and we could increase the sensitivity of the method by about 3 orders
of magnitude as compared to excitation with thermal
light sources.   10 ns are easily achieved with Q-
switch Lasers now-a-days, but with a little more technological effort one would go down to 10 ps if there
is need for it.   In fact people do this though not in
biological samples where the decay times are more

likely to be in the microsecond region. The other
fluorescence parameters one would look at anyway, but
I feel that if you take the native cell and excite it
there will be a large number of molecules fluorescing
and you will be glad about every additional independent parameter which you can vary in order to unfold
the complicated spectra, and this is why I mentioned
it. Another parameter by the way is temperature,
since we are usually working on deep-frozen samples
and have all the possibility for varying the temperature right there. It can be expected that quite a
few of the molecules of interest have a transition
from fluorescence to phosphoresecence in the acces-
sible temperature range.

COLEMAN: I would like to ask both, Dr. Hillenkamp
and Dr. Marich, about the depth of analysis because
the instrument for analysis that Dr. Marich described
is obviously looking from the top down and Dr.
Hillenkamp's is looking from the bottom, and I wonder
what sort of ranges of penetration the laser light
has into the sample and how deep the volume is you
are analysing. I would guess that this would be a
more serious holdback in Dr. Hillenkamp's instrument,
My simple model is that if the sample is very thick
the plasma might remain between the cover slide and
the bulk sample and never get into the mass-spectrometer.

MARICH: We generally penetrate the sample about 1/3
of the diameter of the crater. We generally like to
cut our tissue to a specific thickness so when we
analyze we vaporize a cylinder of tissue.

HILLENKAMP: The limiting quantity in our case is the
focus parameter of the light beam. If we wanted to
go down to 1u sampled area or less we will have to
use numerical apertures of 1 or above. Therefore,
the sampled volume shall be no deeper than its diameter . Otherwise the diameter would increase too.

So if one wants 1μ spatial resolution the section should be no thicker than about 1μ.

BEEUWKES: In each of the techniques the crater size must depend on the absorption of the light. If you go from place to place within the cell you would expect a varying analyzed volume. If you wanted a quantitation, how would you go about to correct for that?

MARICH: In our laser microprobe system we use a Q-switch ruby which we find to be the most reproducible way of sampling. If we take a homogeneously doped material and sample it fifteen or twenty times, we find that the variation of sampling is generally around 5 to 10 percent.

BEEUWKES: But for the cell matrix the homogeneity of the matrix is the point of interest.

HILLENKAMP; The cells do not absorb in the visible, except for a very few components, such as melanine or hemoglobin. There is no usual absorption and vaporization process in the cell. It is a nonlinear process, most probably based on two photon absorption. You have essentially a transparent sample and raise your power level above the threshold for this non-linear processes. One would not expect that this process is influenced by the inhomogeneity of the cells or tissue.

BEEUWKES: Is there any difference from element or matrix to matrix?

HILLENKAMP: Maybe there is one, but very little is known about the details of the plasma formation process even for simple substances, such as solid hydrogen. It is, however, rather unlikely for wavelengths longer than 300 nm.

HALL: I would like to ask Dr. Hillenkamp about the

problems of visualization. In your arrangement you have got a tissue section in the vacuum. If you want to work at them with 1μ or 2μ resolution, I wonder how you are going to obtain images which show sufficient structure at that level so that you can see what you want to see in order to carry out your analysis.

HILLENKAMP: This is one of the reasons why we have chosen the arrangement we have. Usually the laser beam comes from below through the condensor and you are then limited in the type of images you can produce. In our arrangement the laser beam is focussed through the objective and we can easily look and have looked at our samples in phase contrast, interference contrast, light-field and dark-field illumination. The only restriction is that the condensor has to be movable and shifted to the side before the laser is fired.

HALL: But do you feel that this arrangement will give you a sufficiently good image of dry material?

HILLENKAMP: Yes, the resolution of an optical microscope is good enough for the purpose and there are few biological structures which cannot be seen in phase contrast, interference contrast or absorption.

LECHENE: I would like to go back to the question of quantation. Dr. Hillenkamp has not yet looked at biological material and Dr. Marich has shown that the matrix effect of biological material is very great so I come back and ask what is the influence of the matrix effect on the quantation because we know that the cell is an inhomogeneous medium.

HILLENKAMP: We do not, as I said, expect an influence of the matrix on the plasma formation process, however, the different ionization potentials of the various elements do have to be considered. A quantitative measurement will therefore be very difficult and will have to rely on suitable standards as all the other methods do.

A COMPARATIVE STUDY OF MODERN AND FOSSIL MICROBES, USING X-RAY MICROANALYSIS AND CATHODOLUMINESCENCE

M. D. MUIR, L. H. HAMILTON, P. R. GRANT
AND R. A. SPICER
*Department of Geology, Royal School of Mines,
London, England*

INTRODUCTION

One of the most rapidly expanding fields of geological research is the study of fossil micro-organisms. There are many reasons for this expansion of interest, one being that in very ancient rocks (for example of Precambrian age, older than 600×10^6 years) a great number of discoveries have recently been made of microbial remains (Barghoorn and Tyler, 1965; Barghoorn and Schopf, 1966; Schopf, 1968; Engels *et al.*, 1969; Nagy and Nagy, 1969; Schopf and Blacic, 1971; Brooks and Muir, 1971, and many others). The micro-organisms found in the Precambrian show great diversity of form and obviously belong to many different classes of the plant kingdom. They are well preserved in siliceous rocks (e.g. chert), and it is important to determine their affinities to present day microbial groups in order to learn about the evolution of living organisms.

A further reason for the expansion of interest in fossil microbes is that because many of them precipitate, either actively or passively, economically valuable minerals, such as metal sulphides, and metal oxides, it has become of importance to understand the processes leading to the fossilisation of such

33

microbes as an aid to our understanding of the gene-
sis of sedimentary ore.

The present paper discusses how we may assess the
processes involved in the fossilisation of such organisms using X-ray microanalysis, cathodoluminescence
and UV fluorescence techniques.

MATERIALS AND METHODS

Materials and Sample Preparation

Both fossil and present day microbes were examined in this study. Assemblages of present day algae
and bacteria were taken from the lake at the Imperial
College Field Station, at Silwood Park, near Ascot,
Berkshire. The lake waters are turbid and the banks
are stained orange-brown with a heavy growth of *Sphaerotilus*. The samples contained algae belonging mainly to the blue greens (three species of *Oscillatoria*),
with two types of diatom (including *Navicula*), one
Euglenid, some nematodes, *Paramecium*, and many smaller coccoid organisms. The spiral bacterium *Saprospira* was also found. Samples were taken both from
the lake plankton adhering to aquatic vegetation and
from the sediment surface. The algal assemblage dif-
fered markedly from the epiphytic plankton to the
sediment surface. All the above mentioned microbes
occur in the plankton, but in the sediment, only the
iron bacteria, plus the siliceous frustules of the
diatoms, with an occasional blue green algal sheath
were preserved. In artificial experiments, by keep-
ing the plankton samples in a limited light environment, and allowing them to stagnate for periods of up
to three months, the same decrease in variability of
species was observed.

For the microscopic examination, the material was
mounted on glass slides for the UV and light microscope, and on photographic film for X-ray microanalysis

in the SEM. The sheet of photographic film was then mounted on a specimen holder in the normal way, and carbon or aluminum coated. Specimens for cathodoluminescence were mounted on carbon coated aluminum specimen holders, since previous experience had suggested that the aluminum of the specimen holder may emit luminescence in the ultraviolet. Cathodoluminescence specimens were examined uncoated.

The lake samples were examined both unfixed, and fixed in 5% formaldehyde solution.

The fossil material consisted of jaspers (red cherts, composed of a mixture of ferric oxide and silica) from Saudi Arabia and Western Australia. The jaspers were ground into petrological thin sections (30 μm thick) for optical examination. For SEM examination, the rock was fractured to expose a fresh surface, mounted on a specimen holder with conducting paint, and coated with aluminum. Since the X-ray detector on the SEM is a Si(Li) solid state detector, it cannot detect elements below the atomic number of neon, and for carbon analysis, a sample was prepared for an electron probe with crystal (wavelength dispersive) spectrometers.

Because of problems of contamination which would arise from the normal procedures carried out to prepare polished surfaces of material for electron probe microanalysis, e.g. carbon contamination from the grinding powder (silicon carbide), diamond from the cutting wheel, and oil from the lubrication medium, special precautions were taken. The samples were prepared using only alumina powders, and deionised water. The surfaces were finally cleaned ultrasonically, and the material was examined uncoated.

Equipment Used

The light optical microscopy was carried out using a Zeiss Photomicroscope, and the UV microscopy using a Leitz Ortholux microscope with transmitted illumination.

The electron probe microanalysis for carbon was carried out on an electron probe microanalyser, Cambridge Scientific Instruments Microscan V, fitted with wavelength dispersive (crystal) spectrometers. The SEM used was a Cambridge Scientific Instruments Stereoscan Mark IIA, fitted with a solid state Si(Li) X-ray detector. Cathodoluminescence is detected by a second photomultiplier tube fitted at an angle of 45° to the secondary electron detector. The output signal from the cathodoluminescence detector is amplified and fed back into the normal viewing system of the microscope.

Operating Conditions

The SEM was operated at an accelerating potential of 20 kV for microanalysis, and a low incident beam current of about 1×10^{-10}A. Count rates were low, generally of the order of 500 — 800 cps, and in order to obtain statistically valid results, long counting times were used, generally of the order of 1000 seconds.

The electron probe was run at an accelerating potential of 15 kV and an incident beam current of about 1×10^{-6}A in order to achieve adequate count rates for the wavelength dispersive spectrometers. A cold finger in the specimen chamber was used to prevent carbon contamination on the specimen surface which might be caused by this high incident beam current.

The SEM was run at varying accelerating potentials for cathodoluminescence examination, although about 10 kV appears to be the most appropriate, with an incident beam current of about 1×10^{-10}A. It is possible to examine uncoated biological material under such circumstances without excessive charging (Muir

& Grant, 1971).

Methods Used

No account is here necessary of the theory of X-ray microanalysis, since it has been covered in many previous publications (see Birks, 1971; Russ, 1971; Beaman & Isasi, 1972; Lifshin & Cicarelli, 1973, for reviews; and the present volume for problems in biological applications).

The use of cathodoluminescence techniques is less well known, and some account of the theory and practice is therefore necessary.

Materials when excited by incident radiation may emit light either in the visible, UV or IR. The phenomenon has been familiar for many years in the case of UV-stimulated emission. The light emitted by the initial de-excitation of the molecules is frequently referred to as fluorescence. If de-excitation occurs in two stages, the second stage is described as phosphorescence. In principle, there is no reason why the same phenomena should not be excited by an incident electron beam (giving rise to cathodoluminescence — excitation by cathode rays) and it has been shown (Pastrnak and Karel, 1968) that the UV and cathodoluminescence emission of a phosphor are generally identical, both in the emitted wavelengths and in the decay times and rise times of the emission (Figure 1). However, the intensity of cathodoluminescence emission is less than that of the UV emission intensity, and Pastrnak and Karel suggest that this may be because the excited volume is less when the electron beam is used as the excitation medium than for UV, or that electron traps become depopulated by fast electrons in the cathodoluminescence experimental set up.

There is a relatively small volume of work dealing with cathodoluminescence, although it is potentially as useful an analytical method as UV

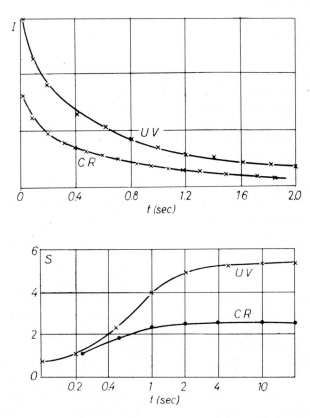

Figure 1. *Showing the correspondence of decay times (top) and rise times (bottom) of UV and cathodoluminescence (CR) of A1N powder activated by different concentrations of manganese (after Pastrnak & Karel, 1968).*

fluorescence, and it may be that one of the reasons for this is the problem of collecting low intensity emission. More efficient collection methods will improve this situation. Most commercially available SEM cathodoluminescence attachments consist of a photomultiplier tube with or without a light-gathering lens in front of it. It has been found that mirrors of various kinds are generally more efficient than lenses, even the type suggested by Davey (1966), and the parabolic (Muir *et al.*, 1971) and elliptical mirrors (Hörl & Mügschl, 1972) both provide a considerable gain in collection efficiency.

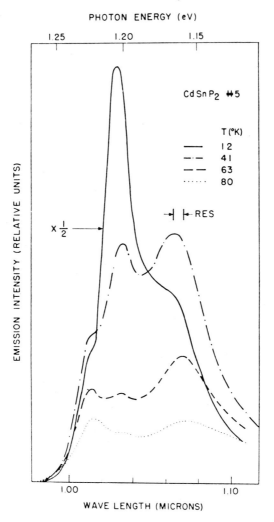

Figure 2. *Cathodoluminescence spectra for a crystal of* CdSnP$_2$ *for various temperatures at a constant beam current of 5 mA. (After Shay et al., 1970).*

Emission efficiency can be considerably improved

by cooling the specimen. Low temperature work at,
say, liquid helium temperatures (Shay *et al.*, 1970)
produces much higher output intensities (Figure 2),
and it has now become feasible to work at these temperatures in the SEM.

Another difficulty in cathodoluminescence studies
is that the detected signal generally has a poor sig-
nal/noise ratio. This is because using commercially
available detection equipment, it is all too easy to
collect back-scattered electrons, and produce a false
cathodoluminescence display. Use of Hörl & Mügschl's
elliptical mirror virtually eliminates this problem.

As in UV-stimulated fluorescence, decay times are
often unfavourably short, again creating problems in
cathodoluminescence detection and display. However,
the potential amount of information available through
cathodoluminescence studies where the spatial resolution of the light emission is much better than that of
the UV microscope is very great, and it seems worthwhile to try to solve some of the problems outlined
above.

RESULTS

Luminescence — Recent Microbes

The light optical examination of the microbes revealed no obvious morphological differences between
the fixed and unfixed material. In the UV microscope,
the blue green algae and *Sphaerotilus* sheaths were
fluorescent, the colour ranging from greenish blue to
whitish. They were also strongly birefringent between
crossed polars. In UV illumination, an apparently
smaller number of sheaths were visible than could be
seen using normal visible illumination, and this phenomenon can be explained by the fact that the majority
of the sheaths are encrusted with oxides and chlorides
of iron. Iron, even in minute traces, quenches

fluorescence, and it seems likely that even when no encrustation is apparent, the iron may be complexed with the mainly polysaccharide material of the sheath wall.

A similar phenomenon was observed in the SEM during the cathodoluminescence examination. In the unfixed material, no luminescence could be detected, because of the quenching effects of the iron. In the formaldehyde fixed material, everything appeared to be luminescent. This result confirms the findings of Manger and Bessis (1970) who observed that red blood cells fixed with paraformaldehyde vapour were strongly luminescent despite their iron content.

Luminescence — Fossil Microbes

No luminescence, either UV or cathodoluminescence, was observed in any of the fossil microbes.

X-Ray Microanalysis — Recent Microbes

The optical appearance of the microbes is shown in Figures 3 and 4. Figure 3 shows a typical association of *Oscillatoria* with *Sphaerotilus* sheaths in a background of colloidal iron hydroxide (unfixed material), and Figure 4 shows a discarded blue green algal sheath in material which has been fixed. Such sheaths are commonly found in Precambrian microbe assemblages.

In general terms, it was found that fixation lowered the concentrations of elements in the preparations (see Table I). This was almost certainly due to the repeated washing in distilled water to which the samples were subjected during the fixation process. The relative concentration of the various major elements in the samples (P, Cl, and Fe) were also affected, however, and Cl is more readily removed than either Fe or P. If extremely long count times were employed, very similar results could be obtained from fixed and unfixed material. Even when the lake bottom

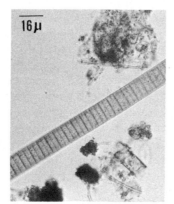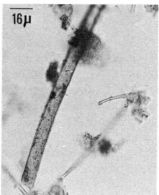

Figures 3-16 are micrographs of present day microbes; Figures 17-28 are of fossil microbes.

Figure 3. *Light optical view of Oscillatoria and Sphaerotilus plus colloidal iron hydroxides, unfixed (BGB2).*

Figure 4. *Light optical view of empty sheaths of Oscillatoria and Sphaerotilus with colloidal iron hydroxides, fixed (BGB5).*

sediment was fixed, the same results were obtained, showing that the colloidal matter present in the lake bottom sediment, and which does not appear biomorphic, is also very soluble in distilled water. Figures 5 and 6 show typical examples of the type of material sampled. Figure 5 is one of the *Oscillatoria* sp. which has been subject to microanalysis for 1000 seconds. The micrograph was taken after the analysis, and the specimen shows some slight damage and it is also charging to a certain extent. Figure 6 shows typical filaments of *Sphaerotilus* and a comparative analysis was carried out on the smooth (unencrusted) forms and on the encrusted specimens (Table II). The unencrusted specimens gave nearly as many Fe $K\alpha$ counts as did the encrusted specimen. This supported the interpretation deduced from the lack of luminescence in even the smooth sheaths that the unencrusted

TABLE I

Fixed (BGA5) and unfixed (BGA1) present day microbes, from the plankton in the lake at Silwood Park, Ascot, Berkshire, England

Sample	Peak	Back-ground	Peak-Background	Peak Background	Peak-Bkgd. Background
Sample BGA1					
P Kα	875	430	445	2.3	1.3
Cl Kα	2050	390	1660	5.3	4.3
Fe Kα	1649	126	1523	13.1	12.1
Sample BGA5					
P Kα	36	9	27	4.0	3.0
Cl Kα	33	14	19	2.4	1.4
Fe Kα	56	6	50	9.3	8.3

TABLE II

Comparison of major elements in a smooth sheath and an encrusted sheath of *Sphaerotilus* from the lake at Silwood Park, Ascot, Berkshire, England

Sample	Peak	Back-ground	Peak-Background	Peak Background	Peak-Bkgd Background
Smooth Sheath					
Si Kα	8178	3066	5112	2.6	1.6
Cl Kα	3246	806	2440	4.03	3.03
Ca Kα	2633	166	2467	15.8	14.8
Fe Kα	1008	133	875	7.5	6.5
Encrusted Sheath					
Si Kα	8255	3021	5234	2.7	1.7
Cl Kα	3575	666	2909	5.4	4.4
Ca Kα	1162	311	851	3.7	2.7
Fe Kα	1403	177	1226	7.8	6.8

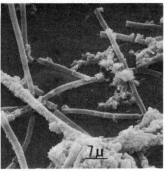

Figure 5. *Secondary emissive mode micrograph of blue-green alga (Oscillatoria) which has undergone microanalysis, unfixed (BGA1).*

Figure 6. *Secondary emissive mode micrograph of Sphaerotilus, with some sheaths which are smooth and some encrusted with iron minerals. Note also the presence of associated colloidal iron hydroxides, unfixed (Sn 3).*

filaments contained iron complexed within their wall structure. Figures 9 and 10 show high power views of the smooth and encrusted filaments of *Sphaerotilus* and *Saprospira*.

Figures 7 and 8 show the secondary emissive mode (Figure 7) and Si Kα distribution micrographs of the diatom *Navicula*. The background of silicon is low, and it seems that most of the silica in the sediment is present in the form of diatom frustules. Figures 11, 12, 13, and 14 show the secondary emissive mode, Cl, Fe, and P distribution micrographs of a cluster of *Sphaerotilus*. It is clear that the iron and chlorine are actually in or on the sheaths, while the phosphorus distribution has no such indication of association with the bacterial sheaths. Table III shows a comparison between the overall composition of sample

 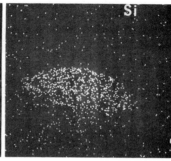

Figure 7. *Secondary emis-*
sive mode micrograph of a
diatom frustule (Navicula)
associated with Oscilla-
toria and Sphaerotilus,
unfixed (BGB2).

Figure 8. *Si Kα distri-*
bution micrograph of the
same field shown in Figure
7.

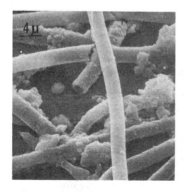 

Figure 9. *Unfixed sheaths*
of Sphaerotilus showing
some smooth and some en-
crusted. Note the extreme
brittleness of the sheaths;
several have fractured dur-
ing preparation (Sn 3).

Figure 10. *General view of*
the plankton (BGA1) (un-
fixed) showing the variety
of sheaths of iron bacter-
ia. Note also the spiral
bacterium Saprospira (cen-
tre of micrograph).

Figure 11. *General view showing unfixed encrusted and unencrusted Sphaerotilus sheaths with colloidal (top of micrograph) and partially crystalline (bottom right hand corner) mineral matter. (Sn 3).*

Figure 12. *Cl Kα distribution of the field of view in Figure 11.*

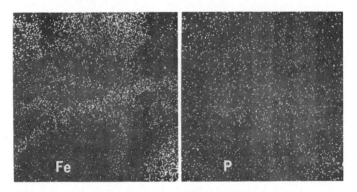

Figure 13. *Fe Kα distribution of the field of view in Figure 11.*

Figure 14. *P Kα distribution of the field of view in Figure 11.*

TABLE III

Comparison of a general analysis of sample BGB2 (planktonic material from the lake at Silwood Park, Ascot, Berkshire, England) with a single blue-green alga (*Oscillatoria*) and a diatom frustule (*Navicula*) from the same material. An analysis of colloidal mineral matter from the same sample is also included for comparative purposes. (Unfixed material).

Sample	General BGB2			Blue-Green			Diatom			Colloidal Mineral Matter		
Element	Peak	Bkgd	P/Bkgd	Peak	Bkgd	P/Bkgd	Peak	Bkgd	P/Bkgd	Peak	Bkgd	P/Bkgd
Si Kα	1781	907	1.8				1306	435	3.0	2555	223	11.4
P Kα				472	241	1.9	674	435	1.5			
S Kα				343	205	1.6						
Cl Kα	4711	998	4.7	567	207	2.7	1185	337	3.5	780	217	3.6
Ca Kα	6791	516	13.1									
Fe Kα	2572	480	5.3							690	69	10.0

BGB2 (unfixed plankton sample) and *Oscillatoria* and a
Navicula from the same sample. Table II indicates
that although the morphological appearance of smooth
and Fe-encrusted *Sphaerotilus* sheaths is very different (see Figures 9, 10, 11 and 15), their major element composition differs in respect of only one element — calcium. The proportion of calcium is very
much higher in the smooth than the encrusted filaments. It seems possible that this calcium causes the
observed birefringence in the optical microscope.
Furthermore Table III shows that the total assemblage
of microbes contains a high proportion of detectable
calcium unlike any of the algal species analysed.
This is clearly associated with the bacterial sheaths
(see Table II).

Finally, Table III indicates the composition of
the colloidal mineral matter (see Figure 16) which
eventually becomes incorporated in the sediment. The
detectable elements consist of Si, Cl, and Fe. Aluminum may also be present, but the peak for this may be

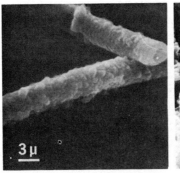 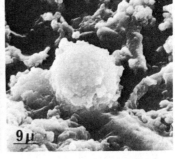

Figure 15. *High power
view of encrusted sheaths
of Sphaerotilus, unfixed
(Sed 4).*

Figure 16. *Colloidal sphe-
roid of iron hydroxides.
(Sed 4).*

spurious and produced by the excitation of the specimen holder.

X-Ray Analysis — The Fossil Microbes

Figures 17, 18 and 20 show iron and silica encrusted filaments from a weathered chert at the Evelyn
Molly Prospect, south of Southern Cross, Western
Australia. The rock is dark red in colour and the
fresh fractured surfaces seen in the SEM are covered
by anastomosing filaments of varying size (see Figure
18) which are encrusted with silica and iron oxides.
The composition of one filament and its matrix is giv-
en in Table IV. There is very little difference in
composition between the filament and the matrix. In
the example in Table IV, the filament appears to be
silica poor with respect to the matrix, while in the
example illustrated in Figures 20 and 21, there is a
decrease in Fe Kα associated with the position of the
filament.

It is clear that morphologically, the modern filaments shown in Figure 15 and the fossil ones shown
in Figure 17 are very similar. Their compositions are
also comparable. In Figures 19 and 22, optical micrographs of jasper from Saudi Arabia show considerable
resemblances with the *Sphaerotilus* micrographs (Figures 3 and 4). Figure 23, also a jasper from Saudi
Arabia, illustrates colonies of very minute filamentous iron bacteria, as well as filaments of the type
and size of *Sphaerotilus* sheaths.

In the modern material, there is clearly a great
deal of organic carbon present in various forms, as
polysaccharides, hydrocarbons etc., and it was necessary to try to establish whether or not it was present
in the fossil material. The X-ray microanalyser attached to the SEM is unable to detect carbon, and the
analyses were carried out on an electron probe with
wavelength dispersive spectrometers. The sample chosen was an onkolite (a laminated organic/sedimentary

Figures 17-28 are micrographs of fossil microbes.

Figure 17. *Fossil fila-*
mentous structures (see
Figure 15) from Southern
Cross, Western Australia
(PM7a3). Preserved in
jasper (a mixture of iron
oxide and silica).

Figure 18. *Fossil filamen-*
tous structures preserved
in jasper from Southern
Cross, Western Australia
(PM7a3). The two filaments
have different diameters —
the one on the left may be
an encrusted fungal hypha.

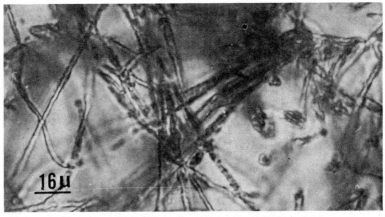

Figure 19. *Petrological thin section (30 μm*
thick) showing filaments of approximately the same
dimensions as the present day Sphaerotilus. Some are
almost smooth; others are noticeably encrusted. From
a jasper vein, Wazra, Saudi Arabia.

Cathodoluminescence Studies

TABLE IV

Comparative analyses of a filament and its rock matrix. Jasper from Southern Cross, Western Australia (PM7a3).

Sample	Peak	Background	Peak-Background	Peak Background	Peak-Bkgd. Background
Filament					
Si Kα	4105	827	3278	4.9	3.9
Cl Kα	1057	541	516	1.9	0.95
Fe Kα	7487	409	7078	18.3	17.3
Matrix					
Si Kα	17458	1623	15835	10.6	9.6
Cl Kα	3252	1772	1480	1.8	0.84
Fe Kα	17532	900	16632	19.5	18.5

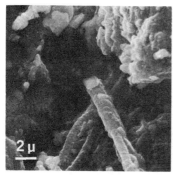 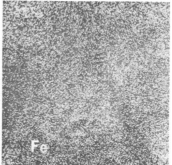

Figure 20. *Filamentous structures from jasper, Southern Cross, Western Australia. Both the filaments are preserved in opaline silica.*

Figure 21. *Fe Kα distribution micrograph of the same field as shown in Figure 20. There are fewer iron counts from the region where the filaments are.*

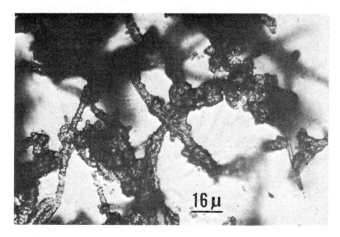

Figure 22. *Petrological thin section of vein-filling jasper from Wadi Fatima, Saudi Arabia. Filaments heavily encrusted with iron hydroxide set in a matrix of chert (crypto-crystalline silica).*

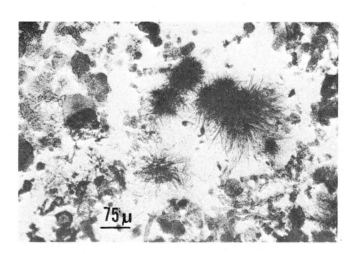

Figure 23. *Low power micrograph of part of the same thin section of jasper. The "whiskery" structures in the centre of the micrograph are composed of large numbers of radiating tubules preserved in iron hydroxide. The overall structure of the colonies resembles that of some present day iron bacteria. Wadi Fatima, Saudi Arabia.*

structure) which had been partly silicified. It contained some residual calcium carbonate, and although it was possible to see that the carbon distribution followed the lamination of the rock (Figures 24 and 25), analyses for calcium and carbon at higher magnification (Figures 26, 27, and 28) show that the distributions for calcium and carbon are quite different. This is interpreted as showing that much of the carbon present in the rock is organic carbon, and not the carbon from calcium carbonate.

DISCUSSION

It is clear from the above that the similarities between present day and fossil microbes are great in terms both of morphology and of chemistry. The cathodoluminescence study first suggested that because of the lack of luminescence in the smooth sheaths of *Sphaerotilus*, there might be iron present in the wall in some form or other. This was confirmed by the X-ray microanalysis data. It was also possible to confirm the results of Manger and Bessis (1970) who showed that fixation with paraformaldehyde caused luminescence.

The X-ray microanalyses indicated that there was organic carbon associated with the fossil microbes, and that compositional changes took place in the present day assemblages as the plankton gradually became incorporated into the lake bottom sediment. The sediment was mainly composed of Si, Fe, and Cl, whereas, among the plankton, P, and Ca were abundant, with some sulphur. The fossil examples were also composed mainly of Fe, Si, and Cl, although in different proportions, and this lends strength to the morphological comparisons that have been made. The fossil microbes were deficient in Ca, as was the lake bottom sediment in the present day examples.

The fossil micro-organisms described in this

Figure 24. *Back-scattered electron micrograph surface of onkolite (finely laminated algal sedimentary structure) showing the concentric banding (Wadi Fatima, Saudi Arabia).*

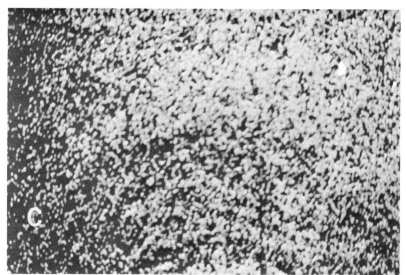

Figure 25. *C Kα distribution over the area shown in Figure 24.*

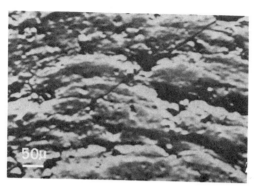

Figure 26. *Back-scattered electron micrograph of parts of the specimen in Figure 24.*

Figure 27. *C Kα distribution of the area shown in Figure 26.*

Figure 28. *Ca Kα distribution of the area shown in Figure 26.*

paper are not particularly old. They are Tertiary in
age (about 60×10^6 years), and in the case of the
Australian specimens at least, the micro-organisms
must have been introduced into the rocks during the
extended weathering cycle to which that continent has
been subjected. Marshall, May and Perret (1966) described some sphaeroidal bodies, made up of oxides
and hydroxides of iron from the Evelyn Molly locality,
and suggested that they might be indigenous to the
rock and hence very old (2.600×10^6 years old). The
filaments described here, however, were discovered in
deliberately selected weathered rocks, and it is possible that the material described by Marshall *et al.*,
is of the same age as the weathering.

ACKNOWLEDGEMENTS

We are indebted to Mr. K. A. Plumb of the Bureau
of Mineral Resources, Canberra, who collected the material from the Evelyn Molly Prospect with one of us
(M. D. M.), and to Mr. H. F. King for his constant
interest and support.

We are also very grateful to Mr. H. Priestly and
Mr. J. Winn of the R.A.R.D.E. at Fort Halstead,
Sevenoaks, Kent, for allowing us to use their Micro-
scan V, and to Mr. P. Suddaby of the Royal School of
Mines, Imperial College, London, for his advice on the
probe analysis.

The cathodoluminescence and X-ray microanalytical
equipment was purchased using funds supplied by the
Natural Environmental Research Council.

REFERENCES

Barghoorn, E. S., and Schopf, J. W., 1965, *Science,*
150, 337-339.

Cathodoluminescence Studies

Barghoorn, E. S., and Tyler, S. A., 1965, *Science*, <u>147</u>, 563-575.

Beaman, D. R., and Isasi, J. A., 1972, *ASTM, STP 506*, *Am. Soc. Test. Mater.*, 1-80.

Birks, L. S., 1971, *Electron Probe Microanalysis*, Wiley-Interscience, New York, 1-190.

Brooks, J., and Muir, M. D., 1971, *Grana*, 9-14.

Davey, J. P., 1966, in *Optique des Rayons X et Microanalyse*, (Eds. Castaing, R., Deschamps, P., and Philibert, J.), Hermann, Paris. 436-441.

Engel, A. E. J., Nagy, B., Nagy, L. A., Engel, C. W., Kremp, G. O. W., and Drew, C. M., 1968, *Science*, <u>161</u>, 1005-1007.

Hörl, H., and Mügschl, M., 1972, *Proc. Vth European Congress on Electron Microscopy*, Manchester, 451-452.

Lifshin, E., and Ciccarelli, M. F., 1973, *Proc. VIth SEM Symposium, IITRI*, Chicago, 89-96.

Manger, W. M., and Bessis, M., 1970, *VIIth Cong. Int. Microscopie Electronique*, Grenoble, 483-486.

Marshall, C. G. A., May, J. W., and Perret, C. J., 1964, *Science*, <u>144</u>, 290-292.

Muir, M. D., and Grant, P. R., 1971, *in* Brooks, J., *et al.*, *Sporopollenin*, Academic Press, London and New York, 422-439.

Muir, M. D., Grant, P. R., Hubbard, G., and Mundell, J., 1971, *Proc. IVth SEM Symposium, IITRI*, Chicago, 401-408.

Nagy, B., and Nagy, L. A., 1969, *Nature,* 223, 1226–1229.

Pastrnak, J., and Karel, F., 1968, *Proc. International Luminescence Conference 1966,* 1473–1476.

Russ, J., (Ed.), 1971, *ASTM, STP 485, Am. Soc. Test. Mater.,* 1–483.

Schopf, J. W., 1968, *Journ. Paleont.,* 42, 651–688.

Schopf, J. W., and Blacic, J. M., 1971, *Journ. Paleont.,* 45, 925–960.

Shay, J. L., Leheny, R. F., Buehler, E., and Wernick, J., 1970, *Journal of Luminescence,* 3, 855–864.

ELEMENTAL ANALYSIS OF BIOLOGICAL SECTIONS USING AN X-RAY FLUORESCENCE MICROPROBE

W. A. P. NICHOLSON
Institute of Medical Physics,
University of Münster, Germany.

INTRODUCTION

The chemical composition of biological specimens has been measured by x-ray emission spectroscopy for a number of years, the two principal methods of generating the spectra being electron and x-ray excitation. Electron excitation in the electron microprobe analyser has the advantage of high spatial resolution, of the order of 1 μm^2, and a high sensitivity in terms of the minimum detectable mass ($\sim 10^{-16}$ g). Although x-ray excitation is not likely to achieve better performance based on these criteria, it has the advantages that the specimen damage will be much smaller and the specimen preparation techniques much simpler than in the electron microprobe. There is no need for the specimen to be covered with a conducting coating, such as aluminium, and some suitable specimens may be analysed wet.

The primary beam from the micro-focus tube has a high specific intensity, and for this reason has been used as the x-ray source in a number of x-ray probe analysers (Zeitz and Baez 1957, Long and Cosslett 1957, Zeitz 1961). The primary beam is generated by a focused beam of electrons which strike a thin foil that is part of the vacuum wall, so that the specimen chamber can remain at atmospheric pressure (Figure 1). The x-ray beam is collimated by an aperture which effectively defines the spatial resolution of the instrument.

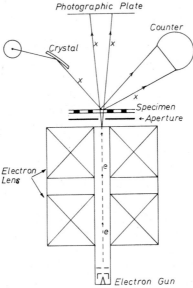

Figure 1. *Principle of the X-ray Fluorescence Analyser.*

The spectrum may be detected by either diffractive or non-diffractive spectroscopy. However, provided that there is no interference between elemental lines, non-diffractive detectors give a much higher sensitivity, mainly because the counter can collect a greater solid angle of x-rays from the specimen (Long and Röckert, 1963).

One difficulty of elemental analysis of thin sections is that both the thickness and the density may vary from one region to another of the specimen, with a corresponding change in x-ray counting rate, so that this quantity will not be a reliable measure of the elemental concentration. Hall (1958; 1961) has pointed out that the intensity of scattering of the characteristic line of the primary spectrum is approximately proportional to the total mass of the specimen in the beam. A simultaneous measurement of this line with the elemental line provides a method of measuring elemental mass fractions almost independant of the specimen thickness.

EXPERIMENTAL

The apparatus described here was designed to maximize the fluorescent intensity by bringing the specimen very close to the x-ray source. Figure 2 shows the arrangement of the apparatus on the second condenser lens of the electron optic column. The electrons could be focused into a spot of about 1 μm on to the target foil, which was usually copper or titanium 3 μm thick. The primary x-ray beam thus generated

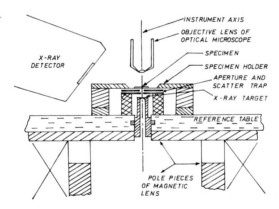

Figure 2. *Experimental arrangement of the analyser. (Nicholson and Hall, 1973, courtesy of the Institute of Physics).*

was collimated by tantalum apertures in the range 50 μm to 200 μm in diameter. Above them a set of similar, slightly larger, apertures were placed to trap the radiation scattered by the edges of the collimating apertures. By adjusting the heights of the specimen and the target, the specimen could be brought within 200 μm of the x-ray source. The region to be analysed, and the aperture beneath it, could be viewed with an optical microscope.

W. A. P. Nicholson

THEORY

X-rays can only lose energy in discrete quanta so that the spectrum generated in the specimen contains only fluorescent lines, and no continuum can be generated. The background detected is continuum radiation from the primary beam, which has been scattered by the specimen or the aperture.

The whole spectrum from the specimen was measured by a silicon detector and two energy channels were selected by pulse height analysis, one the elemental channel E, centered on the fluorescence line, and the other monitor channel M, set in some part of the scattered spectrum. The ratio of the counts in these two channels, after background subtraction can be expressed:

$$r \doteq \frac{F + S_E}{S_M} \qquad (1)$$

where F is the intensity of the fluorescence line and S_E and S_M are the intensities for the spectral regions scattered into two channels E and M respectively (Figure 3). Assuming that a normally incident monochromatic x-ray beam excites the line of a single heavy element in an organic matrix, the fluorescence intensity F is then proportional to the difference of two terms. The first is due to the absorption of the incident line in the specimen and the second is due to the absorption of the emergent fluorescence beam:

$$F \propto C \; (exp^{-\mu_I \, m} - exp^{-\mu_F \, m \, sec \, \phi}$$

in which C is the elemental concentration, μ_I and μ_F are the mass absorption coefficients for the incident and fluorescence wavelengths, m is the mass thickness of the specimen and ϕ is the take-off angle to the detector. Figure 4 shows the computed fluorescence intensity as a function of the specimen thickness for

Analysis Using X-ray Fluorescence

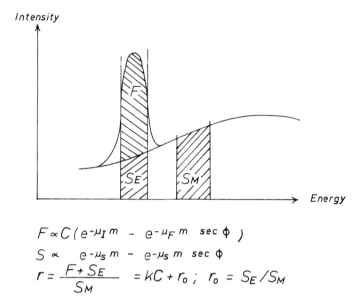

$$F \propto C \left(e^{-\mu_I m} - e^{-\mu_F m \ sec \ \phi} \right)$$

$$S \propto \ e^{-\mu_s m} - e^{-\mu_s m \ sec \ \phi}$$

$$r = \frac{F + S_E}{S_M} \ = kC + r_0 \ ; \ \ r_0 = S_E / S_M$$

Figure 3. *Principle of measurements in two channels.*

several concentrations of potassium in an organic matrix of sucrose. It is evident that unless the specimen thickness were accurately known, it would not be possible to determine the concentration from a simple measure of line intensity.

The intensity scattered from a band of the continuum S, is expressed by a similar function to the fluorescence intensity but in this case the mass absorption coefficients μ_S are the same in both terms as the same wavelength enters and leaves the specimen:

$$S \ \alpha \ exp^{-\mu_S \ m} - \ exp^{-\mu_S \ m \ sec \ \phi}$$

Figure 5 shows the ratio r, of the fluorescent to scattered intensities, and it can be seen that the ratio is nearly independant of specimen thickness. This curve, which is the calibration curve for the element, shows that equation 1 can almost be expressed

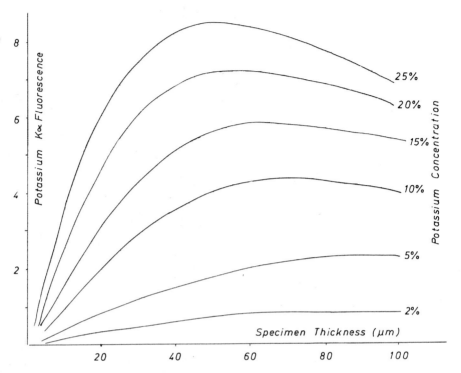

Figure 4. *Computed fluorescent intensity for potassium in an organic matrix.*

as the simple equation of a straight line:

$$r = kC + r_0$$

where k is approximately constant and $r_0 = S_E/S_M$ is the ratio of the intensities that would be scattered by a specimen of pure organic matrix into the two channels. Ideally k and r_0 are constants independant of specimen thickness. In practice this is best approximated when the incident, fluorescence and monitor wavelengths are situated close together so that their absorption in the specimen is similar.

For high efficiency of fluorescence generation the incident x-ray must be of just higher energy than

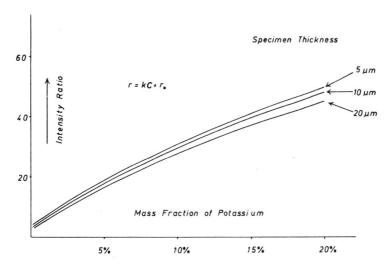

Figure 5. *Computed ratio of fluorescent to scattered intensities.*

TABLE I

Summary of the Measurements on Sucrose Standards

Element	Excited by	Calibration Constant k	Limit of Measurability p.p.m.
Potassium	Continuum	340	80
Manganese	Copper Kα	980	140
Sulphur	Continuum	68	2000
Sulphur	Titanium Kα	900	60
Potassium	Titanium Kα	1300	35

k is the constant in the calibration equation

$$r = kC + r_0$$

the absorption edge of the excited element. For example, titanium Kα (4.5 keV), would be suitable for the generation of potassium Kα (3.3 keV), and this could also be selected as the monitor wavelength, thus satisfying the condition that all three energies (or wavelengths) are similar.

RESULTS

To investigate the performance of the instrument and also to prepare calibration curves, a set of standards was made of spots of sucrose about 50 μm thick containing the element of interest in mass fractions in the range 0.02% to 30%. The diameter of the primary beam was collimated to about 100 μm so that the total mass irradiated was about 1 μg. A number of elements were measured, with the spectral lines being generated by the radiation from copper and titanium targets. The monitor channel was set in a region of the continuum 600 eV wide, placed so as to avoid any spectral lines, so that the efficiences of fluorescence generation, indicated by the value of k, could be directly compared in each case. The summary of the results, shown in Table I, suggests that for high efficiency, the atomic number of the target should not be more than six higher than that of the measured element, in this region of the periodic table.

Figure 6 shows a typical calibration curve, in this case for potassium. The curve is used in estimating r_0 and Δr_0, the probable error in r_0. For the smallest measurable concentrations, r approached r_0 and since the error in measuring k was relatively small:

$$C \ (min) \sim \frac{n\sqrt{2}}{k} \Delta r_0$$

where $1/n$ is the fractional error in $C \ (min)$. The error in r_0 is the combination of the statistical error in the measured counts, and the uncertainty in r_0 due to variations in specimen thickness. The quantity Δr_0 was estimated from a linear plot of the lowest mass fractions measured, and used to calculate the lowest measurable concentrations shown in Table I.

To exemplify the operation of the instrument when used for analyses other than those of standards, a

series of measurements was made of the potassium concentration in odontoblasts taken from rats' teeth.

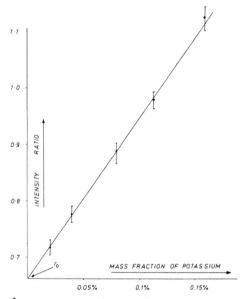

Figure 6. *Measured calibration for potassium.*

The cells were removed in clumps from the inside of
the tooth and placed directly onto the specimen support film, where they were allowed to dry in a clean
atmosphere. These specimens are somewhat thicker
than normal sections, but the values shown in Table
II illustrate the results obtained and data manipula-
tion required. As the matrix of the specimen, soft
tissue, differs from that of the sucrose standards
the results must be corrected by multiplying by a
factor which is equal to the ratio of the scattering
cross sections of the two materials for the monitor
wavelength.

FUTURE PERFORMANCE

The resolution of the fluorescence analyser is
determined by the diameter of the aperture collimating

TABLE II

Results of Odontoblast Measurements

Specimen	Measured Counts		Background Subtracted		Ratio r	Percentage Potassium (uncorrected)
	Monitor Channel	Potassium Channel	Monitor Channel	Potassium Channel		
Background	230	470				
1	332	1,150	102	680	6.8	1.6
2	966	9,200	736	8,730	11.4	2.8
3	385	1,855	155	1,385	8.8	2.1
4	980	7,170	750	6,700	8.9	2.1

Integration time for measured counts: 40 sec.
Standards: Potassium carbonate in sucrose 50 μm thick.
Calibration: $k = 340$ $r_0 = 0.75$.

the primary x-ray beam. The difficulty of manufacturing good apertures has limited the resolution of the present instrument to 50 μm. It is, however, interesting to estimate the possible performance of the fluorescence analyser for much smaller areas. For large areas the smallest mass detectable is limited by the inability to distinguish between the fluorescence counts and the scattered counts in the elemental channel. The presence of the background radiation scattered by the aperture has relatively little effect. For smaller apertures, the lowest measurable mass fraction will be much greater, and will be limited by the extraneous background above which the fluorescence signal must be detected, and the radiation scattered by the specimen will be less important.

To evaluate the limit of detectability as a function of the aperture size, the fluorescent and the scattered counts were taken to be proportional to the mass of the specimen in the beam and hence to the square of the aperture diameter, and the background counts, which derive principally from the edges of the aperture, were assumed to be proportional to the aperture diameter. Figure 7 shows the smallest mass and mass fraction that could be expected to be measured in a 10 μm thick specimen in an integration time of 10 min with a proportional error of 1/2.

It is thought that some improvement of future performance over the values shown in Figure 7 is possible. The x-ray apertures used had, on a microscopic scale, quite rough edges since they were manufactured by mechanical drilling. Apertures made by chemical etching or electron beam machining would be much smoother, thus reducing the background intensity which is the principal limiting factor for small apertures. The x-ray detector had a resolution of 250 eV (full width half height at Mn Kα) compared with 160 eV available at the present time. Since to a first approximation, the background in the elemental channel will be proportional to the channel width, a higher

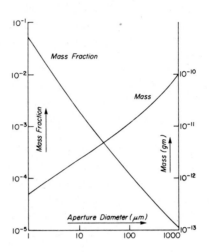

Figure 7. *Estimated limit of measurable potas-ium in a 10 μm section (Nicholson and Hall, 1973, courtesy of the Institute of Physics).*

resolution detector would permit the elemental channel to be of a narrower energy band. This would improve the signal to noise ratio in the channel for the same number of elemental counts, lowering the limit of detectability.

A further possibility to improve the overall performance would be to use a multichannel analyser with a computor. The background subtraction would be more accurate, so that the background in the energy channel of the measured line could be used to monitor the specimen mass, thus making the absorption for the monitor and fluorescence lines equal, thus reducing the error in r_0 due to variations in specimen thickness.

To conclude, it is useful to review the relative advantages of the electron microprobe and the x-ray microprobe. As stated in the introduction, the x-ray probe can rival neither the spatial resolution nor minimum detectable mass of the electron probe. However the problem of specimen damage caused by the electron beam is not yet fully resolved. Current measurements

(Hall, this symposium), show that about 30% of the specimen mass is lost practically instantaneously. Cooling the specimen helps reduce the damage but does not prevent it entirely. The damage to the specimen is comparatively negligible in the x-ray probe.

To detect very small masses, small specimen volumes must be excited. For example, it has been estimated (Hall, 1971), that at a beam diameter of 30 nm, 10^{-18}g could possibly be detected by electron excitation. Conversely, to measure very small concentrations larger volumes must be excited. In the x-ray probe this may be achieved by increasing both the aperture diameter and the specimen thickness. It is believed that by this approach non-destructive measurements in the 10 p.p.m. range could be realized.

REFERENCES

Hall, T. A., 1958, In: *Advances in X-ray Analysis*, Plenum Press, New York, 1, 297-305.

Hall, T. A., 1961, *Science*, 134, 449-455.

Hall, T. A., 1971, In: *Physical Techniques in Biological Research*, Academic Press, New York, 1A, 2nd Ed. 203-206.

Long, J. V. P., and Cosslett, V. E., 1957, In: *X-ray Microscopy and Microradiography*, Academic Press, New York, 435-442.

Long, J. V. P., and Röckert, H. O. E., 1963, In: *X-ray Optics and Microanalysis*, Academic Press, New York, 513-521.

Nicholson, W. A. P., and Hall, T. A., 1973, *J. Phys.E. (Sci. Instr.)*, 6, 781-784

Zeitz, L., 1961, *Rev. Sci. Instr.*, 32, 1423-1424.

Zeitz, L., and Baez, A. V., 1957, In: *X-ray Microscopy and Microradiography*, Academic, New York, 417-434.

W. A. P. Nicholson

DISCUSSION

SCHIMMEL: What targets did you use?

NICHOLSON: I used thin copper and titanium foils.
Although it is most efficient if the primary line of
the target matches the absorption edge of the measured
element, in many cases there is no suitable metal. In
such cases you can only take a metal of high melting
point and thermal conductivity, so that you can put
in as much energy as possible, without burning through
the target.

RUSS: What sort of power levels were you working at?

NICHOLSON: The instrument was usually operated at 20
to 25 kV beam voltage with target currents in the
range 6-10 μA.

HEINRICH: What range of wavelengths would you
consider suitable to measure in that system? I would
think that sodium K would be somewhat difficult?

NICHOLSON: Of course that depends on the detector.
With our present Si(Li) detector sodium is measureable.
I don't think one could go much lower in the fluorescence instrument as the absorption in the specimen in
the transmission geometry is quite high.

HEINRICH: Also the fluorescence yield will go down.

KAUFMANN: Do you see any possibility to improve this
system to a point where it can really be compared with
the electron microprobe?

NICHOLSON: Well, as I said, the main problem is in
the manufacture of the apertures, and the curve I
gave was rather speculative. But if good apertures
can be made then I think the resolution may be improved to 3 μm. Certainly there is not much hope of
anything better. As far as mass fraction is concerned,

Analysis Using X-ray Fluoresence

the sensitivity can go down below the limit of the electron probe if one is prepared to measure slightly larger areas.

HILLENKAMP: There are, in principle, other ways to focus X-rays.

NICHOLSON: I discussed the possibilities of using torroidal mirrors with the people who did the original work, but I don't think they achieved any gain in primary intensity.

HILLENKAMP: I'd like to draw your attention to another possibility. One can get spatial resolution by having good resolution in the excitation, but of course you can alternatively try to use an imaging system on the emission side. As the method is non-destructive, you could illuminate a larger area and use a Fresnel zone plate for imaging the surface. This method is possible in principle, and the resolution can be of a few micrometres.

KAUFMANN: I think in X-ray astronomy people have developed Fresnel zone plates for wavelengths down to carbon K. Does anyone know of any extension of this work to the wavelength of sodium, for example?

RUSS: It seems to me that you have the possibility of making another type of microprobe instrument at little additional cost. If the detector were placed over the specimen to measure the transmitted X-rays, then you could do absorption work. With 10-100 μm thick sections, you should get some useful information.

NICHOLSON: Zeitz has shown that, theoretically at least, fluorescence should be more sensitive than absorption, but the instrument could certainly be modified to make absorption measurements.

ELECTRON AND ION PROBE MICROANALYSIS
PHYSICAL BASES

KURT F. J. HEINRICH

*Institute for Materials Research, National
Bureau of Standards, Washington, D. C.   20234*

INTRODUCTION

Probe techniques applied to biological tissue
sections are frequently advocated as a device to ascertain the topographic distribution of certain elements. The use of laser (Peppers *et al.*, 1968) and
x-ray microprobes (Long and Röckert, 1963) has been
advocated for this purpose. Neither technique requires the exposure of the specimen to vacuum. However, considerations of spatial resolution usually
favor the use of the electron, and more recently, the
ion probe (Figure 1).

A discussion of these techniques must concern itself with the range of observable elements and their
detection limits, with the spatial characteristics of
the probe-specimen interaction, with limits to quantitation, with the conservation or destruction of the
specimen during analysis, and with the techniques of
specimen preparation and orientation for the analysis.

RANGE OF ELEMENTS, SENSITIVITY

The detection and determination of elements with
the *electron probe microanalyzer* is based on the emission of characteristic x-rays by the specimen under
bombardment with electrons. Hydrogen and helium can

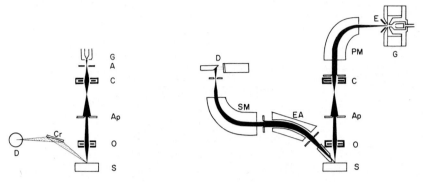

Figure 1. *Schematics of the electron probe (left) and ion probe (right). G = gun, A = anode plate, E = ion extractor, C = condenser lens, O = objective lens, Ap = aperture, S = specimen, PM = primary magnet, EA = electrostatic analyzer, SM = secondary magnet, Cr = analyzer crystal, D = detector. The primary magnet is omitted in some instruments.*

not be observed since they emit no characteristic lines or bands; moreover, the observation at trace levels of all elements of atomic number up to 10 or 11 is difficult or impossible. Elements of atomic number above 11 emit radiations of the K, L, and/or M series which are reasonably easy to detect and measure; the sensitivity is highest for atomic number 13-30 (Al to Zn), and decreases for heavier elements. Absolute limits of detection are in the order of $10^{-13} - 10^{-15}$ g., corresponding to relative limits in the order of $10 - 1000$ ppm. In the conventional sense, electron probe analysis is thus not a trace method. The poor thermal conductivity and stability of soft tissue tend to further increase the limits of detection. As a consequence, the average concentrations in soft tissue of most elements are below the limit of detection of the electron probe microanalyzer. However, local concentrations are frequently, in cases of practical interest, higher by several powers of ten, and can therefore be observed and measured.

In the *ion probe* microanalyzer, the signals used

for the observation of elements are the secondary ion
emissions which are analyzed with a mass spectrometer.
All elements can, in principle, be observed. The sensitivity is, in general, highest for elements of low
atomic number.

The limits of detection in ion probe microanalysis depend on the ratios of the lines of the mass
spectrum to the background. These ratios vary considerably with the resolution of the secondary mass spectrometer — which in some instruments is below the
present state-of-the-art in mass spectrometry, with
the yield of ions — which changes very widely with
element and matrix, with the nature and settings of
the detector system, and with the working conditions
(primary ion species, polarity of primary and secondary ions, ion beam current, and energy). Very high
line-to-background ratios are commonly achieved, and
limits of detection in the ppm or even ppb range are
predicted from these ratios. However, actual experimental evidence of such sensitivities, particularly
in organic matrices, must still be produced. In mass
spectrometers which cannot distinguish organic fragments from elemental ions of the same mass number, the
interpretation of very small peaks must be made very
carefully. Notwithstanding these caveats, the analyst
can expect significant advantages from the use of the
ion probe in the observation of traces of elements.

This expectation is raised still higher by the
possibility of observing isotope ratios in the ion
microprobe. The introduction of compounds labeled
with stable isotopes into biological systems may produce indications of the pathways of the labeled molecules and radicals down to a subcellular level. If
this avenue proves fruitful, the significance of biological ion probe analysis will transcend that of mere
elemental microanalysis, and we will have a tool of
the greatest promise for the study of biochemical mechanisms on the trace level. But again, accurate isotope ratio measurements of elements present at low

concentrations may require spectrometers of better resolution than presently available on ion probes.

SPATIAL LIMITATIONS

Probe analysis relates spectrography to micro-scopy; unless the information requested for a speci-men concerns local levels of concentrations, probes are not appropriate tools for the task. For this rea-son, the discussion of the spatial resolution is of prime importance. In this respect we must distinguish the lateral width of the zone of signal emission, and the depth of emergence of the signal (range, see Fig-ure 2).

In electron probe analysis, the energies of the primary electrons determine their range, z_r, within the specimen, according to the equation (Andersen and Hasler, 1965):

$$z_r(\mu m) \simeq \frac{0.064}{\rho\,(g/cm^3)} \ (E_0^{1.68} - E_q^{1.68})\ (kV) \qquad (1)$$

in which ρ is the density of the specimen, and E_0 and E_q are, respectivly, the operating potential and the critical excitation potential for the excited line. In most cases, particularly in organic matrices, the loss of x-rays emerging from depths less than this range, while not negligible for quantitation, is not significant to the spatial definition of the analysis. Assuming for dry soft tissue a density of 0.42 g/cm^3, we obtain from equation (1) the values for z_r shown in Figure 3.

The effect of E_q is seen to be minor at the over-voltage $(E_0 \geq 2E_q)$ required to obtain acceptable x-ray intensities; the x-ray generation increases roughly with the same power as the range. There is thus, for trace analysis in materials of low density, a conflict between the requirements for resolution in depth, and those for signal intensity.

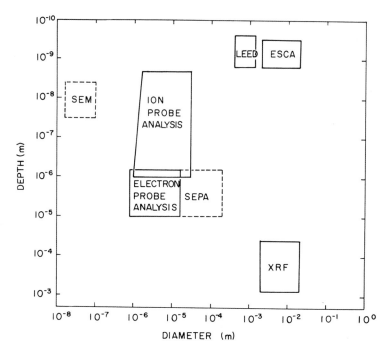

Figure 2. *Spatial resolution of diverse micro-analytical techniques. The dimensions of the graph are, on a logarithmic scale, the depth (vertical) and the diameter (horizontal) of the volume emitting a signal. SEM = scanning electron microscopy; SEPA = scanning electron probe microanalysis; ESCA = electron spectroscopy for chemical analysis; LEED = low-energy electron diffraction; XRF = x-ray fluorescence spectrography.*

The following steps may be taken to reduce the depth of penetration:

a) The specimen can be presented to the beam in the frozen, hydrated condition, provided that by cooling and other protective steps the sublimation of ice and the deposition of ice from the atmosphere in preliminary steps can be prevented. Or, water can be displaced by an interstitial material. Such steps will, of course, effectively diminish the concentration of the elements to be measured, with respect to

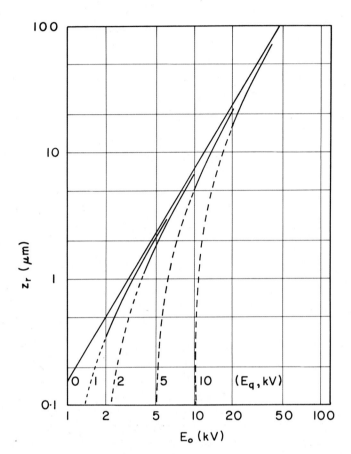

Figure 3. *Depth-range of x-ray emission from dry soft tissue (ρ = 0.42 g/cm³), according to equation (1). E_o = operating voltage (kV), E_q = critical excitation potential (kV).*

the dehydrated specimen.

    b). The specimen can be sectioned. In such cases, it is advantageous to work at high electron beam energies. This technique was followed in Duncumb's Electron Microscope-Microanalyzer (EMMA) (Cooke and Duncumb, 1966), and is equally applicable to scanning transmission electron microscopy techniques involving x-ray detection.

Electron and Ion Probe Microanalysis

The thin section can, depending on the problem, be frozen, dehydrated, or cut from an embedded specimen. The techniques of preserving a frozen, not dehydrated specimen are, of course, very demanding, but they may be the only way to show movable ions at their original sites.

The problems of sensitivity for trace detection become even more acute with the use of thin sections, particularly when a high lateral resolution is also required. The maximum current achievable with a given instrument and voltage varies with the power 8/3 times the beam diameter (Castaing, 1960). The efficiency of x-ray detection can be considerably improved, however, by using an energy-dispersive lithium-drifted silicon detector (Fitzgerald *et al.*, 1968) at close distance to the specimen. This solution is particularly attractive for soft tissue specimens which have simpler spectra than most mineralized materials.

In the electron probe microanalyzer, the electron beam has typically a diameter of .2-.5 μm, while this dimension is reduced in the scanning electron microscope to 100-200 Å. However, due to scattering of the electrons, the width of the excited area within a thick (semi-infinite) specimen may be considerably larger than the beam diameter, being roughly equal to the depth range of analysis (Figure 4). In the thin specimen, the beam broadening is much less significant, as the probability of scattering acts is greatly reduced. Hence, the use of thin specimens improves the lateral resolution as well as the resolution in depth, and is mandatory for the study of minute specimen regions.

In the ion probe, the secondary ions are emitted from the first few atomic layers of the specimen. Therefore, secondary ion spectrometry can be used as a means for virtually non-destructive surface analysis (Benninghoven, 1971). This, however, requires the irradiation of a relatively large area. Even more so

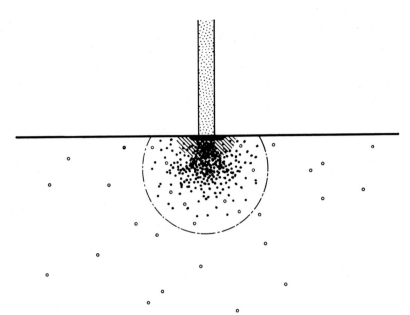

Figure 4. *Regions of signal emission caused by the impact of an electron beam. Broken circle: range of primary electrons. Black: region of emission of secondary electrons. Hatched: region of emission of backscattered electrons. Black circles: sites of generation of primary x-ray photons. Open circles: sites of generation of fluorescent x-ray photons.*

than with the electron probe, the spherical aberration of the beam-forming optics limits the achievable intensity of a beam of 1-2 μm diameter. Furthermore, a finely focused beam of the intensity necessary for the detection of ppm levels erodes the site of analysis quite rapidly. Clearly, a compromise must be taken between spatial resolution and sensitivity of detection, and the extremely low limits of detection frequently cited are not achievable with a beam of 1-2 μm diameter. In practice, this limitation is frequently unimportant when the study of distribution in depth (over relatively large surfaces) is of interest. This, in the biological domain, could be applied to concentration variations across membranes if appropriate specimen preparation procedures were used.

It should also be taken into account that the mass spectrometers now used in ion probes provide information on only one mass number at a time. If one is interested in variations in the ratio of two emitted ion species, one must use techniques in which the signals are observed before the specimen erosion or instability of the ion source have altered their ratio. In this context it is important that the stability of presently used ion sources is inferior to that of a conventional electron source.

QUANTITATION

In principle, the prospects of achieving accurate analyses are much brighter in electron probe analysis than in ion probe analysis. A respectable body of theory and procedures has been accumulated, and several computer programs for quantitative data evaluation are being used (Beaman and Isasi, 1970; Hénoc et al., 1973). The emission of x-rays is caused by transitions in the energy levels of the orbital electrons, and therefore x-ray spectra can be fairly accurately related to the mass fractions of the elements in the specimen. At present, errors not exceeding 2-3% relative (for major constituents) can be expected in inorganic specimens; the non-destructive nature of electron probe analysis allows repetition of the measurement at the same spot.

However, scientists working in biological media have, in general, a very limited interest in the "classical" theories of electron probe microanalysis (Hall, 1968), and for very good reasons. As mentioned above, the large depth of penetration of electrons usually forces the biologist to operate with thin specimen sections, to which the classical theory is not well applicable. Due to the structure of biological tissue, its local density varies greatly. In

specimen preparation techniques which dehydrate the
specimen, the local distribution of mobile ions, such
as Na+ and K+, may be profoundly altered. If, as in
conventional histological techniques, the tissue water
is displaced by other substances, further alterations
of composition are possible. Finally, there is evidence of decomposition of biological tissue irradiated by an electron beam, even if provisions were made
to protect the specimen from the heat produced on
beam impact and from contamination by the beam. The
carbon and metal coatings applied to the specimen for
protection further complicate the situation. The combination of these facts renders quantitative analysis
of soft tissue by the conventional techniques practically impossible.

Approaches to quantitation of thin biological
specimens will be discussed by other authors. I will
mention, however, that techniques based on Monte-Carlo
calculations are increasingly used for thin specimens,
and may contribute to reduce some of the limitations
which presently plague electron probe analysis of biological specimens.

In these calculations, the paths of individual
electrons are obtained by a sequence of applications
of random numbers, weighted according to the probabilities of the respective interactions. The parameters to take into account are the deceleration of
electrons — usually approximated by Bethe's law —,
the free path between elastic scattering acts, and the
total and azimuthal angles of scattering. When an
electron recrosses the boundaries of the specimen,
it is considered transmitted or backscattered, depending on its direction. Once a statistically meaningful number of trajectories has been obtained, the
parameters of interest, such as emitted and absorbed
x-ray intensities, electron backscatter and transmission coefficients, etc., can be obtained as averages. The great advantage of the method lies in the
flexibility with which the boundaries of the specimen

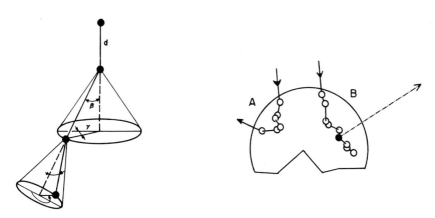

Figure 5. *Paths of electrons in the specimen.*
Left: spatial parameters — d = path between colli-
sions, β = scattering angle, γ = azimuthal angle.
Right: two typical paths of penetration, terminating
in backscattering (A) and x-ray generation (B).

can be defined (Figure 5).

The quantitative aspects of ion probe analysis
are at present less known than those of electron probe
analysis. Both the sputtering yield for the specimen
atoms and their ionization yield vary greatly for different elements; the resulting secondary ion intensities range over several orders of magnitude, and are
strongly subject to matrix effects (Andersen, 1969).
The absolute signal intensities are also dependent on
the orientation of the surface, and in crystalline
materials channeling effects may produce variations as
well. Most investigators rely on relations of intensities of various lines. Andersen has established a
model for data reduction which is based on thermody-
namic functions (Andersen, 1969) which appears to be
useful in the analysis for minor constituents of minerals, particularly if the concentrations of major
constituents are known. A test of application of such
corrective schemes to the analysis of soft biological
tissue would require the use of homogeneous standards.

As the composition of biological materials does not fluctuate to as large an extent as that of minerals and alloys, there is a distinct possibility of quantitation with the help of empirical standards. The fact that soft tissue is, from the point of view of ion penetration, close to the amorphous state, should be another advantage. A problem may arise, however, due to the low electrical conductivity of biological tissues. The functioning of the ion microprobe depends on a strict control of the electrostatic potentials in the area between the specimen and the entrance of the secondary mass spectrometer. If this control is upset, the observed intensities of ion emission vary. It is possible to shield the specimen from electrostatic charging by applying a thin conductive coating which can be perforated by the beam at the points to be analyzed. It is also very desirable to use a negative primary beam for the analysis of insulating materials since, in such a mode of operation, the impact of negative charge can be balanced by the emission of secondary electrons.

A considerable problem may arise for either probe technique from the need to recognize the histological structure or feature to be analyzed. For this purpose, the microscopist uses specific staining methods which serve to identify certain cells or organelles. Such staining techniques, however, cannot be applied if movable ions are to be investigated. Scanning transmission microscopy of thin tissue sections can be employed in electron probe analyzers, but the orientation on frozen solid blocks, or in fact on most specimens during ion probe analysis, presents difficulties, the seriousness of which should not be underestimated.

REFERENCES

Andersen, C. A., 1969, *Internat. J. of Mass Spectrometry and Ion Physics,* 2, 61.

Electron and Ion Probe Microanalysis

Andersen, C. A., and Hasler, M. F., 1965, *in X-Ray Optics and Microanalysis*, Castaing, R., Deschamps, P., and Philibert, J., Eds., Hermann Press, Paris, 310-327.

Beaman, D. R., and Isasi, J. A., 1970, *Anal. Chem.*, 42, 1540-1548.

Benninghoven, A., 1971, *Surface Science*, 28, 541-562.

Castaing, R., 1960, *in Advances in Electronics and Electron Physics XIII*, Marton, L., Ed., Academic Press, New York, 317-386.

Cooke, C. J., and Duncumb, P., 1966, *in X-Ray Optics and Microanalysis*, Castaing, R., Deschamps, P., and Philibert, J., Eds., Hermann, Paris, 467-476.

Fitzgerald, R., Keil, K., and Heinrich, K. F. J., 1968, *Science*, 159, 528-529.

Hall, T., 1968, *in Quantitative Electron Probe Microanalysis*, U.S. Natl. Bur. Stds. Spec. Publ., 298, U.S. Government Printing Office, Washington, D. C., 269-299.

Hénoc, J., Heinrich, K. F. J., and Myklebust, R. L., 1973, *U.S. Natl. Bur. Stds. Tech. Note*, 542, U.S. Government Printing Office, Washington, D.C.

Long, J. V. P. and Röckert, H., 1963, *in X-Ray Optics and X-Ray Microanalysis*, Pattee, H. H., Cosslett, V. E., and Engström, A., Eds., Academic Press, New York, 513-521.

Peppers, N. A., Scribner, E. J., Alterton, L. E., Honey, R. C., Beatrice, E. S., Harding-Barlow, I., Rosan, R. C., and Glick, D., 1968, *Anal. Chem.*, 40, 1178-1182.

CELLULAR MICROANALYSIS: A COMPARISON BETWEEN ELECTRON MICROPROBE AND SECONDARY ION EMISSION MICROANALYSIS

P. GALLE

Départment Biophysique, Faculté de Médecine, Créteil, France; and Laboratoire de Physique des Solides [1] *, Université Paris, France*

INTRODUCTION

A new microanalysis method has been proposed in 1962 by Castaing and Slodzian. This method, called "Secondary Ion Emission Microanalysis", can be used to study biological cells and tissues and can be compared with electron probe microanalysis. Results obtinaed by each of these methods are discussed from the point of view of qualitative analysis, quantitative analysis, degree of sensitivity, correspondence of the analysis with intracellular structures, and resolution. A consideration of these factors and the nature of the specific problem under investigation will allow us to choose the most suitable method.

METHODS

The electron microanalyzer we used is the CAMECA probe. With this instrument, specimens can be observed either with a light microscope, for thick sections, or a transmission electron microscope, for ultrathin sections. The secondary ion microanalyzer is an ion emission microscope combined with a mass spectrometer

---

[1] Action specifique n° A 6580061 du C.N.R.S.

(Castaing and Slodzian, 1962a). With this instrument,
all elements can be analyzed and ion distribution images are obtained directly without the need of any
scanning. Since the technique of secondary ion emission microanalysis is less known than electron probe
microanalysis, a brief review of the physical principles involved is included.

Theory of Secondary Ion Emission Microanalysis

The flat surface of a solid specimen is bombarded
with a beam of primary ions (protons, argon, oxygen or
nitrogen) having an energy of approximately 10 KeV.
As a result of the ion bombardment, the atoms of the
most superficial layers of the specimen are etched.
Some of them remain electrically neutral whereas
others become ionized. These ionized particles, called secondary ions, are a sampling of the elements present at the surface of this specimen. The secondary
ions are accelerated by a potential of 4.4 KeV and
focused into a beam by means of electrostatic lenses.
The beam transports the ion images of all the ions
etched from the surface of the specimen. In order to
form the individual images of the different ion species, a mass spectrometer is used to split the initial
beam into as many secondary beams as there are ions
of a given specific charge (e/m = charge per unit
mass). The mass spectrometer can focuse the beam in
two planes, radial and transversal. It has been ar-
ranged in such a way that the quality of the final
ion image of each of the separated secondary beams
will not be destroyed. Therefore, by a judicious
choice of magnetic field, it is possible to select a
beam containing only ions of a specific charge, and to
form an image which represents the distribution of a
single ion species. To visualize the image thus obtained, another device, an image converter, is necessary which transforms the ion image into an electron
image, since fluorescent screens are not utilizable
directly in this case because of their very weak response to ions. In summary, by changing the settings

of the mass spectrometer one can observe successively the distribution of different ions on the surface of the specimen.

Figure 1. *The C.A.M.E.C.A. Secondary Ion Emission Microanalyzer.*

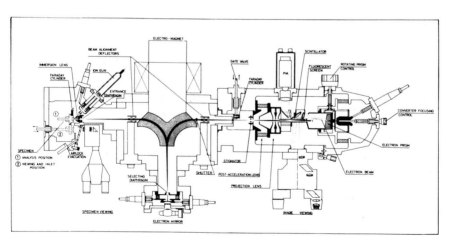

Figure 2. *Schematic representation of the principal parts of the Secondary Ion Emission Microanalyzer showing the path of the beam through the instrument.*

91

The secondary ion emission analyzer used in the solid state physics department in Orsay is shown in Figure 1. Figure 2 is a schematic representation of the operation of the instrument.

Remarks on Ion Optics. All the etched ions do not leave the surface of the specimen with the same initial velocity and in order to maximize the resolving power of the mass spectral analysis the ions are "filtered", first as a function of their momentum (mv) and second as a function of their energy ($\frac{1}{2}$ mv^2). The initial selection is made by means of a magnetic prism. The filtered beam is then aimed at an electrostatic mirror adjusted to reflect only those ions the speed of which is less than a predetermined threshold. In this way very fast-moving ions are eliminated. The magnetic prism and electrostatic mirror are positioned as shown in Figure 3. A more detailed description of this arrangement has been given elsewhere (Castaing, 1965).

Other Characteristics. The vacuum in the specimen chamber is of the order of 5×10^{-8} torr. Under normal working conditions the primary beam is about 0.3 mm in diameter and has a current density of 20 μA/mm^2.

Because of the double filtering system the mass resolving power of the instrument is sufficient for the analysis of all elements of the Periodic Table. The separating power, however, generally is not good enough to show mass defects and to distinguish between monoatomic ions and polyatomic ions having the same mass unit. The resolution of the final image obtained is of the order of 1 μ and may even approch 0.5 μ in the central region.

Special Problems Posed by Biological Specimens

The analysis of biological specimens poses two major problems: the first results from the fact that

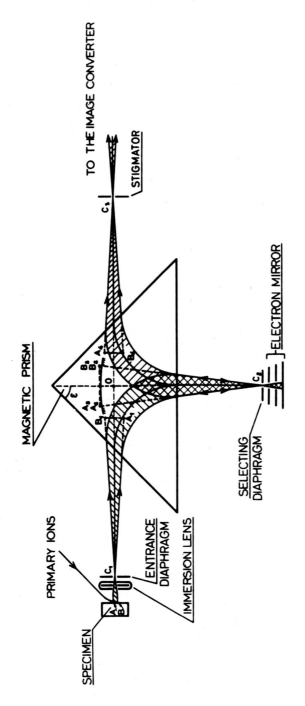

MOMENTUM AND ENERGY FILTERING SYSTEM

Figure 3. *Schematic representation of the function of the magnetic prism.*

they are non-conductors, and the second involves de-
vising methods of preparation which will preserve, as
far as possible, the conditions existing in the living
state.

Conductivity of the Specimen. Theoretically,
secondary ion emission microanalysis can only be performed on specimens which are electrical conductors
and which have a perfectly flat, equipotential surface.
Surfaces of non-conducting specimens may become irreg-
ularly charged during bombardment with the primary ion
beam, with a resultant loss of the image. For this
reason, one of the procedures often used in the study
of nonconductors is the vacuum evaporation of a metallic grid on to the surface of the specimen (Slodzian
1963). The layer of evaporated metal is a few microns
thick with grid bars 0.003 mm wide and grid spaces
0.150 mm on each side. The presence of the metallic
grid allows the free flow of electric charge over the
surface of the specimen. Our experience has shown,
however, that the addition of a metallic grid is not
always necessary, especially in the case where thin
histologic sections (approximately 1 μ) on a conducting support are examined. In this case, the tissue
section acts like a sheet with parallel surfaces and
the support like one of the plates of a condenser
charged to a given potential. The two faces of the
section (especially the free surface which will be
used for the analysis) correspond to an equipotential
surface, charges can flow along the support and in
most cases the analysis can be carried out without
difficulty. We have analyzed sections of 1 μ thickness and 4 mm^2 surface area placed on a carefully
polished stainless steel support. The images obtained were stable. The analysis of thicker sections has
not been possible.

Specimen Preparation. The specimen must be prepared so that the *in vivo* distribution and concentration of the elements present in the different cellular
and extracellular compartments is not changed. This

94

condition is normally not satisfied by the classical techniques of fixation and embedding. The most powerful fixatives, such as OsO_4, fix proteins and lipids at their normal sites but soluble ionized elements (Na^+, K^+, Cl^-, etc.) remain free to diffuse throughout the tissue. Alcohol dehydration and routine embedding also present the same serious drawback. The only valid method at the present is quick freezing (quenching) followed by vacuum dehydration. This is the same difficulty which is encountered in any analysis of diffusable ions (whether by secondary ion emission or electron probe microanalysis) and which has not as yet been satisfactorily resolved. On the other hand, the study of non-diffusible ions does not pose the same problem and here, routine Epon or paraffin embedded sections can be used.

RESULTS

Studies on a Smear of Frog Nucleated Red Blood Cells

After spreading on a perfectly polished metallic support (in this case, gold), the red blood cells were distributed over the surface and separated from each other by a layer of plasma. The ion emission image of $^{23}Na^+$ shows at first very bright areas which correspond to the plasma and cell nuclei. The cytoplasm appears very much darker (Figure 4). As primary bombardment continues, the plasma areas become dimmer and disappear completely in about 3 minutes, leaving only the image of the cells. The cell cytoplasm and nucleus are etched at different rates, the nuclei disappear in general before the cytoplasm (Figure 5), the image of which persists for approximately 15 minutes under normal conditions. Figures 4-8 show the succession of images seen during the progressive etching of the specimen. The fact that the rate of etching varies for the different cellular and extracellular compartments must be taken into consideration in any attempt at quantitative analysis.

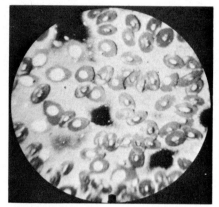

Figure 4. *After 30 sec.*

Figure 5. *After 1 min.*

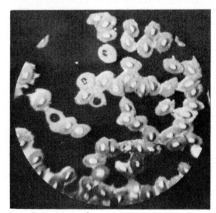

Figure 6. *After 2 min.*

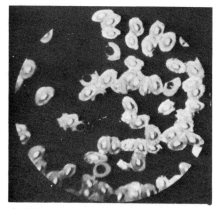

Figure 7. *After 10 min.*

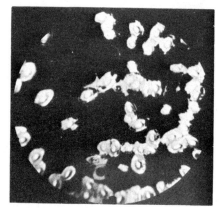

Figure 8. *After 15 min.*

Figures 4 - 8. *Distribution of $^{23}Na^+$ on the surface of a frog blood cell smear. (After the begining of Ion Bombardement). Ion Bombardment: Argon.*

For example, the greater brightness of the areas corresponding to the plasma in the analysis of $^{23}Na^+$ is the result of both a higher concentration of sodium and a more rapid etching of the plasma regions by the primary beam.

Figure 9 shows the distribution of iron ($^{56}Fe^+$) in the hemoglobin of frog blood cells. It also shows, inadvertantly, the difficulty of obtaining metallic supports which contain no impurities. It should be noted here that, at mass unit 56, there is a possible interference with CaO^+.

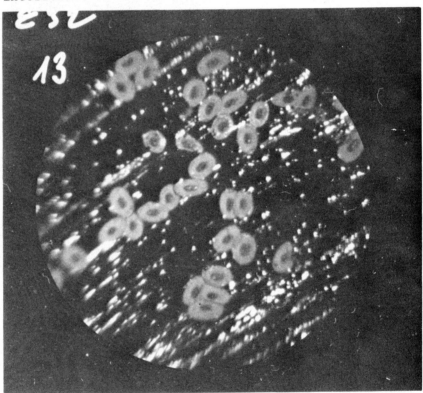

Figure 9. *Distribution of $^{56}Fe^+$ on the surface of a frog blood cell smear. (4 min after the beginning of primary Ion bombardment). Ion bombardment: Argon.*

Electron and Ion Analysis of Cells

Studies on Tissue Sections

Sections of kidney and thyroid were studied by secondary ion emission analysis. The tissue was quick frozen in isopentane cooled by liquid nitrogen. Sections were cut with a freezing microtome at −20°C and freeze-dried under vacuum.

As was seen with the blood smear, the rate of etching varies for the different compartments of the

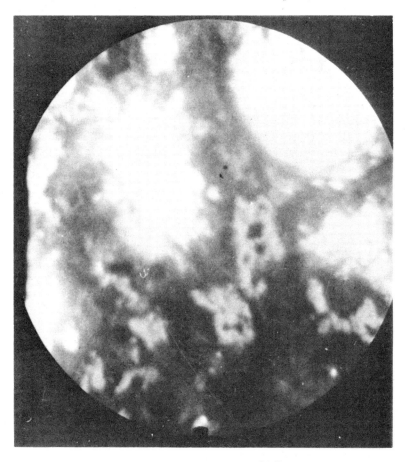

Figure 10. *Distribution of* $^{127}I-$ *on the surface of a rat thyroid section. Ion bombardment: Argon.*

tissue sections, it is very rapid for the extracellular compartments and less rapid for the intracellular (nucleus and cytoplasm). The results show that a certain amount of redistribution of soluble ions occured in spite of the precautions taken in the preparation of the specimen. Interpretation of the images produced by secondary ion emission will only have real significance when the problem of soluble ion diffusion during specimen preparation has been solved. Successful analysis of diffusible ions by both electron probe microanalysis and autoradiography also depends on the solution of this problem. However, when it has been solved and satisfactory sections obtained, secondary ion emission microanalysis may well become the method of choice for these studies because of its very high sensitivity.

Figure 10 shows the distribution of iodine ($^{127}I-$) at the surface of a section of rat thyroid and is an example of the images currently obtained from these preparations.

DISCUSSION

Comparison of the Characteristics of Electron Probe and Secondary Ion Emission Microanalysis for Biological Studies

Qualitative Analysis

Electron Microprobe Analysis. The interpretation of the characteristic X-ray spectra obtained is very simple and therefore a negligible source of error in qualitative analysis. However, elements with atomic numbers less than 5 are not analyzable and isotopes cannot be distinguished from each other.

Secondary Ion Emission Analysis. By this method both stable and radioactive isotopes can be distinguished and analyzed separately. All elements of the

Electron and Ion Analysis of Cells

Periodic Table can be detected and there is no limitation for very light elements. On the other hand, the interpretation of the spectra obtained by mass spectroscopy is sometimes very difficult for the two following reasons:

a) secondary ions of the same element may carry several unit charges (for example, Al^+, Al^{++}, Al^{+++}), and

b) secondary ions may be emitted in a polyatomic form (for example, Al^+, and $C_2H_3^+$ both have mass unit 27; Fe^+ and CaO^+ both have mass unit 56). Because of the possibility of producing polyatomic ions, great care must be taken in the analysis of results from chemically complex specimens such as tissue sections. It is, in fact, very probable that in the course of primary bombardment of cells and their surroundings a significant number of ionized organic radicals are produced. These radicals, composed of several atoms, could easily have the same mass as a given ion or any other element being studied. Systematic studies of the secondary ion emission spectra of organic compounds will be necessary in order to arrive at a better interpretation of results.

Quantitative Analysis

Electron Microprobe Analysis. There is no longer any doubt about the reliability and precision of quantitative electron probe analysis. Nevertheless it should be noted that, with a few exceptions, there have been few biological applications. This fact stems from two principal causes:

a) the intracellular volumes of interest (mitochondria, cytoplasmic vesicles, cytoplasmic areas in the vicinity of certain organelles) are for the most part too small (less than 0.1 μ^3) and b) the local concentrations of the elements of interest are generally too low for the conclusive interpretation of the results obtained.

P. Galle

Secondary Ion Emission Analysis. The difficulties of obtaining a reliable quantitative analysis are as great using this method as with electron probe analysis, but for different reasons. As was described above, the purpose of the image converter is to transform the beam of secondary ions which is carrying the image into a beam of electrons with the same spatial distribution of information, capable of interacting with a fluorescent screen or photographic plate.

Theoretically, it is possible to infer from the current density per unit of surface area (or degree of blackening of the photographic plate) the local concentration of a particular element. However, for the particular case of biological specimens, the rate of etching varies greatly according to the structure to be analyzed. Therefore, in order to quantify local concentrations, it is necessary to measure both the local emission and the relative sputtering speed. Such a measurement may require the use of apertures permitting the selection of very small areas of the specimen under bombardment for analysis.

An absolute quantitative interpretation is further hindered by an imperfect knowledge of the phenomenon of secondary ion emission itself. For example, the intensity of secondary ion emission does not depend exclusively on the concentration of the element, but also on its chemical bonds to other elements in the sample. As a result, the quantitative interpretation of results is a very delicate problem. On the positive side, the fact that secondary ion emission depends on the chemical bonds in which an element is involved, may in certain cases provide information on the nature of these bonds.

Sensitivity

Electron Microprobe Analysis. At the present time, the sensitivity of this method is limited to 10^{-16} or 10^{-17} grams for many elements.

Electron and Ion Analysis of Cells

Secondary Ion Emission Analysis. This method has a sensitivity which is much greater than that of any other previously used method. The sensitivity, however, varies under the different conditions and depends on several factors: the nature of the element itself, its chemical state, and the primary ions used. These different factors determine the ratio N_1/N_0; that is, the ratio between the number of characteristic ions which contribute to the formation of the image (N_1) and the number of neutral atoms which are etched at the same time from the same region of the specimen (N_0). It is evident that the sensitivity increases directly with the value of the ratio N_1/N_0. For sodium, for example, the ratio is probably very close to 1. Using a sample of red blood cells, in which the sodium concentration is well known and evenly distributed, it is possible to calculate the minimum quantity of sodium detectible in 1 μ^2 of cell surface. These experiments show that less than 10^{-19} grams of sodium per μ^2 of red blood cell surface can be detected and photographed. On the whole, the sensitivity is very high for light elements and for those of columns I and II of the Periodic Table. For the other elements, the sensitivity varies greatly. It is very good, for example, for pure aluminium where Slodzian finds a value of 10^{-4} for N_1/N_0, while it is much lower for pure copper where a value of 10^{-7} has been found.

Correspondence Between Intracellular Structures and Analytical Results

This aspect is of central importance in the case of biological applications where the relations between structure, function and chemical composition are continually being sought.

Electron Microprobe Analysis. This method has a great advantage in that the instrument can be associated with a transmission or scanning electron microscope, thus allowing simultaneous observation of the

103

specimen at the ultrastructural level.

Secondary Ion Emission Analysis. Observation
of the specimen at the ultrastructural level is not
possible, unless suitable electron optics are included in the instrument.

Resolution

Electron Microprobe Analysis. Ultrathin sections
of biological material can be studied only by this
method. For thick sections of large pieces of tissue,
resolution depends on the diameter of the probe and
the diffusion of electrons in the specimen. In biological material, the volume of diffusion is fairly
large and it is difficult to obtain a resolution as
good as 1 μ. Moreover, when the concentration of an
element is low, the limit of resolution is much worse
than 1 μ.

Secondary Ion Emission Analysis. At the present
time, the limit of resolution in the image obtained
from the surface of the preparation is of the order
of 0.5 μ at the center and 1.0 μ in the peripheral
areas. Since primary ions penetrate less than 100 Å
into the specimen, the volume of material being analyzed at any given time is much less than that excited
by an electron probe.

CONCLUSION

Two methods of microanalysis have been compared.
One of them, electron microprobe analysis, is older,
well known, and has been tested in many laboratories.
The other, secondary ion emission analysis, is just
beginning to be explored. It is probable that electron
probe analysis, which has already achieved a very high
level of technical advancement, will evolve little;
however, the possibilities of secondary ion emission
analysis await future development.

Electron and Ion Analysis of Cells

REFERENCES

Castaing, R., 1965, IVe Congrés International sur
l'Optique des rayons X et la Microanalyse, 1 vol.,
48, Hermann, Paris.

Castaing, R., and Slodzian, G., 1962a, *J. de
Microscopie*, 1, 395.

Castaing, R. and Slodzian, G., 1962b, *Compt. Rend.*,
Acad. Sc., 255, 1893.

Galle, P., 1970, *Ann. Phys. Biol. et Méd.*, 84, 94.

Galle, P., Blaise, G., and Slodzian, G., 1969, IV.
Nat. Conf. on Electron Microprobe Analysis,
1 vol., 36, A.A. Chodos Ed., California Inst.
of Techn.

Slodzian, G., 1963, *Thesis Sc.* Paris.

Tousimis, A. J., 1972, VII Nat. Conf. on Electron
Probe Microanalysis, 1 vol., 45, California Inst.
Techn.

PRECIPITATION TECHNIQUES AS A MEANS FOR INTRACELLULAR ION LOCALIZATION BY USE OF ELECTRON PROBE ANALYSIS

A. LÄUCHLI*, R. STELZER*, R. GUGGENHEIM**, AND L. HENNING***

*Fachbereich Biologie-Botanik Technische
Hochschule Darmstadt, Darmstadt, Germany
**Geologisch-paläontologisches Institut
der Universität Basel, Switzerland
***Raster-Elektronenmikroskopie-Labor,
Basel, Switzerland

INTRODUCTION

Biologists concerned with problems of ion transport have recently focused on means for intracellular ion localization as an aid in studying ion transport in cells and tissues. During specimen preparation the possibility of displacement or leaching of ions is a serious obstacle to the intracellular ion localization. Such redistribution of ions can be avoided when the ion, which is to be localized, is immobilized *in situ* by a precipitation reaction (cf. Läuchli, 1972a).

Most precipitation techniques are based on the reaction of an ion with a heavy metal producing an electron dense deposit which is visible in the transmission electron microscope. (For precipitation of ions *in situ* by organic reagents see Constantin *et al.*, 1965; Podolsky *et al.*, 1970; Spurr, 1972). Komnick (1962) was the first to use this approach for Cl^- localization in animal tissues by precipitation with organic Ag^+ salts. Subsequently, he tested this

method extensively on animal tissues (Komnick and
Komnick, 1963; Komnick and Bierther, 1969; van Lennep
and Komnick, 1971), and, more recently, it was also
applied to plant specimens (Ziegler and Lüttge, 1967;
Van Steveninck and Chenoweth, 1972; Van Steveninck
et al., 1973). Other ions which may be localized intracellularly by precipitation reactions are Na^+ as
pyroantimonate (Komnick, 1962; Levering and Thomson,
1972) and PO_4^{3-} as Pb-salt (Tandler and Solari, 1969).
While the reactions *in situ* between Cl^- and Ag^+ and
between PO_4^{3-} and Pb^{2+} are considered to be specific,
the specificity of pyroantimonate for Na^+ is question-
able (Lane and Martin, 1969; Tandler *et al.*, 1970;
Weavers, 1971; Yarom and Meiri, 1973). In fact, pyroantimonate may even be bound in tissue by organic
solutes (Winborn *et al.*, 1972).

The elemental composition of such deposits is
obtained by use of electron probe analysis. Thus,
studies on ion localization with transmission electron
microscopy should be supplemented by electron probe
analysis. In this report, evidence is presented on
the validity of precipitating Cl^- in plant tissues
with Ag^+. By means of energy dispersive x-ray analysis (Russ, 1971), it is confirmed that the electron
dense deposits resulting from treatment of Ag^+ contain
Ag^+ and Cl^-. Furthermore, it is shown that the scanning electron microscope fitted with an energy dispersive detector is a versatile instrument for rapid
electron probe analysis of plant specimens (cf.
Läuchli, 1973).

MATERIAL AND METHODS

Experiments were done with seedlings of barley,
Hordeum vulgare, cv. "Kocherperle". Barley seeds were
germinated for 24 hr and then grown for 2 days in the
dark on an aerated 0.2 mM solution of $CaSO_4$. There-
after, the seedlings were transferred to an aerated,
1/10 concentration Johnson nutrient solution (Johnson

et al., 1957); Fe-EDTA, however, was supplied at ½
concentration. The seedlings were grown for 11 days
in the light at the following conditions: 16 hr pho-
toperiod, ca. 10,000 lux light intensity, ca. 30°C
day temperature and ca. 25°C night temperature. Sodium chloride between 0 and 50 mM was either added to
the renewed nutrient solution after 7 days of growth
in the light or NaCl was supplied as a short pulse up
to 30 min at the end of the growth period. The roots
still attached to their shoots were then rinsed for
1 min with ice-cold 0.2 mM $CaSO_4$.

Small root segments cut about 1 cm behind the tip
were fixed for 2 to 3 hr at +2°C with 1% OsO_4, containing either 0.5% Ag-acetate or 1% Ag-lactate in
0.1 M cacodylate-acetate buffer pH 6.5 under red
light. The fixed specimens were washed for 15 min
with cacodylate-acetate buffer and dehydrated continuously with acetone. During dehydration, a brief wash
with 0.05 N HNO_3 in 30% acetone was included to dissolve unspecific Ag^+ precipitates (cf. Komnick and
Bierther, 1969). The dehydrated specimens were transferred into propylene oxide and embedded in Spurr's
epoxy resin embedding medium (Spurr, 1969).

Ultrathin transverse sections for transmission
electron microscopy were mounted on Formvar-coated
Cu-grids and viewed unstained in a Zeiss EM 9 electron
microscope. For analysis with the scanning electron
microscope, 500 nm-thick transverse sections were also
mounted on Cu-grids, the grids in turn being affixed
on carbon discs with silver paint and vacuum-coated
with carbon.

Energy dispersive x-ray analysis was accomplished with a scanning electron microscope (Cambridge
Steroscan Mark 2A) fitted with the EDAX system, model
707, at 20 KeV accelerating voltage and a beam diameter of approximately 200 Å.

To test the possible loss of Cl^- from the

specimen during the preparation procedure, roots of
intact barley seedlings, grown in the absence of NaCl,
were exposed for 30 min to nutrient solution containing 5 mM NaCl, radioactively labeled with ^{36}Cl. The
roots were washed for 1 min with ice-cold 0.2 mM $CaSO_4$,
blotted and excised. Root segments were fixed and
dehydrated as described above, and the fixed specimens as well as the combined fixation and dehydration
fluids radioassayed for ^{36}Cl. It was found that only
a negligible fraction of the total ^{36}Cl absorbed,
i.e. less than 4%, was lost from the specimens during
preparation.

RESULTS AND DISCUSSION

Transmission electron micrographs revealed that
distinct silver deposits occured in root cells when
barley seedlings were grown on nutrient solution containing 5 to 50 mM NaCl and Cl⁻ was precipitated *in
situ* with Ag⁺. Furthermore, these deposits were virtually absent both in sections from plants grown in
nutrient solution lacking Cl⁻, but fixed in the presence of Ag⁺, and in sections from plants grown in
NaCl containing nutrient solution but fixed with OsO_4
alone. Figure 1 shows a transmission electron micrograph of a xylem parenchyma cell from a root grown in
the presence of 50 mM NaCl. Large silver deposits
appear as electron dense spots in the cytoplasm and
the vacuole. The cell wall free-space from which Cl⁻
was exchanged before fixation does not contain any
deposits. Furthermore, silver deposits are practically absent in the mitochondria. When roots were
examined which were grown in the presence of NaCl concentrations lower than 50 mM, the silver deposits were
often much smaller and allowed a more detailed ultrastructural localization of Cl⁻; these results will be
described elsewhere (Stelzer *et al.*, 1973).

The silver deposits are also visible in the scanning electron microscope as bright spots both in

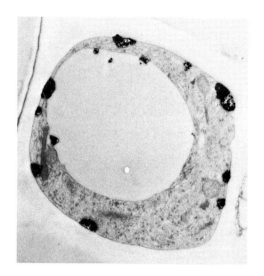

Figure 1. *Transmission electron micrograph of an ultrathin section through a root of a 14-day-old barley seedling grown in nutrient solution which contained 50 mM NaCl for the last 4 days of growth. Precipitation of Cl⁻ in situ with Ag-lactate. Xylem parenchyma cell with large silver deposits in cytoplasm and vacuole. x 4,820.*

Figure 2. *Secondary electron image from a 500 nm transverse section through a root of a 14-day-old barley seedling grown in nutrient solution which contained 25 mM NaCl for the last 4 days of growth. Precipitation of Cl⁻ in situ with Ag-lactate. Parts of 3 cortical cells visible with vacuoles (top, bottom, and left side) and intercellular space (center area). Silver deposits appear as bright spots. x 6,300.*

secondary and backscattered electron images. Figure 2 shows a secondary electron image depicting parts of three cortical cells with a large intercellular space in the center area. Silver deposits are abundant in the cytoplasm and occur also in the vacuoles; the intercellular space contains only few deposits. The cell walls are not clearly distinguishable. Energy spectra were obtained from a silver deposit in the cytoplasm and from a spot of a deposit-free area in

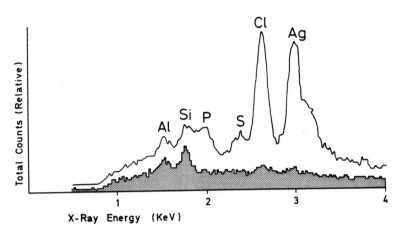

Figure 3. *Energy dispersive x-ray spectra from a silver deposit in the cytoplasm (top) and from a spot of a deposit-free area in the vacuole (bottom) of a cortical cell shown in Figure 2. Counting times 240 sec.*

the vacuole (Figure 3). The spectrum from the silver deposit (top) clearly shows the predominance of Ag and Cl. There is also some P and S present, probably originating from cytoplasm. The Si and Al peaks are probably due to contamination from the scanning electron microscope (cf. Sutfin *et al.*, 1971). The bottom spectrum is a measure for background x-ray radiation from a spot with no silver deposit present. There is very little Ag and Cl, but the Si and Al peaks are again clearly discernible.

The epoxy resin embedding medium used in these experiments contains some Cl (Läuchli *et al.*, 1970). However, this Cl contamination of the embedding medium apparently does not influence the elemental analysis of the silver deposits (Figure 3). A Cl contamination peak was only detectable when pure embedding medium without tissue present was analyzed with the EDAX system, as is demonstrated in Figure 4 (the Cu peak in Figure 4 is due to the underlying Cu grid).

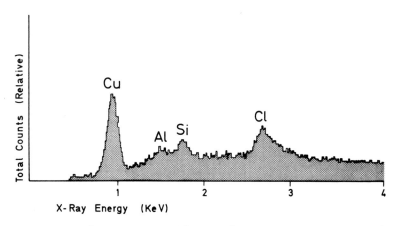

Figure 4. *Energy dispersive x-ray spectrum from a spot in the embedding medium outside the root tissue. Counting time 240 sec.*

NaCl concentrations of 25 and 50 mM in the medium are still within the physiological range of concentrations for many plant species including barley. However, distinct silver deposits were visible even in cells of roots which were exposed to a short pulse of only 5 mM NaCl. Figure 5 shows that after a pulse of 20 min duration the elements of Ag and Cl were detectable in the silver deposits of xylem parenchyma cells.

The problem of whether the intracellular distribution of Cl^- is preserved by the method used in this study is critical. In other words, is the distribution of silver deposits a real image of Cl^- distribution in the intact cell? One could argue that Os affects membranes and may increase their permeability to Cl^-, leading to a loss of Cl^- from the tissue. However, practically no loss of ^{36}Cl to the fixation and dehydration fluids was observed from tissue whose intracellular Cl^- had been labeled with ^{36}Cl (cf. last paragraph in "Material and Methods"). Yet, the action of Os on membranes might cause an altered intracellular Cl^- distribution, if Ag^+ does not rapidly penetrate the specimen. Silver deposits were detectable in cytoplasm and vacuole (Figure 1); therefore, Ag^+

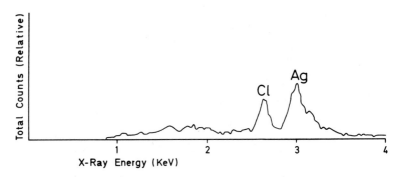

Figure 5. *Energy dispersive x-ray spectrum from a silver deposit in the cytoplasm of a xylem parenchyma cell. Chloride supplied to the root of a 14-day-old seedling as NaCl (5 mM, 20 min). Precipitation of Cl⁻ in situ with Ag-lactate. Counting time 240 sec.*

penetrated both boundary membranes of the cytoplasm, i.e. plasmalemma and tonoplast. In fact, the presence of silver deposits inside the vacuolar membrane means that Ag^+ moved through the cytoplasmic layer into the vacuole more rapidly than vacuolar Cl^- could move out.

In situ precipitation of Cl^- with Ag^+ was tried also in the presence of other fixation media, though without success. Glutaraldehyde cannot be used, because Ag^+ in the fixation medium is reduced immediately to metallic Ag. Potassium permanganate as an alternative oxidative fixation agent was also unsatisfactory, because of poor structural preservation. However, it is planned to verify the pattern of Cl^- distribution obtained after precipitation with Ag^+ by using methods of specimen preparation which do not involve chemical fixation, that is, cryo-ultramicrotomy (cf. Gahan *et al.*, 1970; Appleton, 1972; Echlin, 1973; Gullasch and Kaufmann, 1973) and freeze-substitution (cf. Läuchli *et al.*, 1970; Läuchli, 1972a).

114

Precipitation for Ion Localization

At this time, it seems reasonable to examine whether the intracellular Cl^- distribution obtained with this method is compatible with other physiological and biochemical data. There is good agreement with regard to the following intracellular compartments:

a) ion accumulation in plant cells is thought to be located to a considerable extent in the vacuoles (see e.g. Laties, 1969; Läuchli and Epstein, 1971). Since Cl^- accumulates in plant tissues under the experimental conditions used in this study, the presence of silver deposits in the vacuole (Figure 1) was expected.

b) Mitochondria were almost devoid of silver deposits (Figure 1). Studies with isolated plant mitochondria also showed that these organelles are impermeable to Cl^- (Phillips and Williams, 1973) and do not commonly accumulate this ion (Hanson and Hodges, 1967).

c) In experiments to be published (Stelzer *et al.*, 1973), Cl^- after *in situ* precipitation with Ag^+ was found to be also localized within plasmodesmata. The bulk of physiological and structural evidence on intercellular ion transport in plant roots point very strongly to the existence of symplasmic transport by way of plasmodesmata (see e.g. Laties, 1969; Läuchli, 1972b).
Thus, it appears that the distribution of Cl^- in cells of plant roots after *in situ* precipitation with Ag^+ is in good agreement with other experimental evidence.

CONCLUSION

The Ag^+ precipitation technique for intracellular Cl^- localization is a valid method for studies on Cl^- transport in plant roots, provided that elemental analysis of the silver deposits by electron probe analysis shows the presence of Cl in the deposits. Yet, this precipitation technique may not be specific for Cl^- in certain other biological tissues because of other possible sources of error, e.g. unspecific Ag^+ reduction by cellular components with no Cl^- present. Precipitation techniques for intracellular localization of other ions by transmission electron microscopy should also be checked by electron probe analysis to assure specificity of the precipitating agent for the particular ion.

REFERENCES

Appleton, T. C., 1972, *Micron*, <u>3</u>, 101–105.

Constantin, L. L., Franzini-Armstrong, C., and Podolsky, R. J., 1965, *Science*, <u>147</u>, 158–160.

Echlin, P., 1973, This Symposium.

Gahan, P. B., Greenoak, G. C., and James, D., 1970, *Histochemie*, <u>24</u>, 230–235.

Gullasch, J., and Kaufmann, R., 1973, This Symposium.

Hanson, J. B., and Hodges, T. K., 1967, <u>in</u> *Current Topics in Bioenergetics*, Vol. <u>2</u>, Sanadi, D. R., Ed., Academic Press, New York, 65–98.

Johnson, C. M., Stout, P. R., Broyer, T. C., and Carlton, A. B., 1957, *Plant Soil*, <u>8</u>, 337–353.

Komnick H., 1962, *Protoplasma*, <u>55</u>, 414–418.

Komnick, H., and Bierther, M., 1969, *Histochemie*, <u>18</u>, 337–362.

Komnick, H., and Komnick U., 1963, *Z. Zellforsch.*, <u>60</u>, 163–203.

Lane, B. P., and Martin, E., 1969, *J. Histochem. Cytochem.*, <u>17</u>, 102–106.

Precipitation for Ion Localization

Laties, G. G., 1969, *Ann. Rev. Plant Physiol.*, 20, 89-116.

Läuchli, A., 1972a, in *Microautoradiography and Electron Probe Analysis: Their Application to Plant Physiology*, Lüttge, U., Ed., Springer-Verlag, Berlin, 191-236.

Läuchli, A., 1972b, *Ann. Rev. Plant Physiol.*, 23, 197-218.

Läuchli, A., 1973, in *Liverpool Workshop on Ion Transport in Plants*, Anderson, W. P., Ed., Academic Press, London, 1-10.

Läuchli, A., and Epstein, E., 1971, *Plant Physiol.*, 48, 111-117.

Läuchli, A., Spurr, A. R., and Wittkopp, R. W., 1970, *Planta*, 95, 341-350.

Lennep, E. W. Van, and Komnick H., 1971, *Cytobiologie*, 3, 137-151.

Levering, C. A., and Thomson, W. W., 1972, in *30th Ann. Proc. Electron Microscopy Soc. Amer.*, Houston, Arceneaux, C. J., Ed., Claitor's Publ. Div., Baton Rouge.

Phillips, M. L., and Williams, G. R., 1973, *Plant Physiol.*, 51, 667-670.

Podolsky, R. J., Hall, T., and Hatchett, S. L., 1970, *J. Cell Biol.*, 44, 699-702.

Russ, J. C., 1971, *Energy Dispersion X-Ray Analysis: X-Ray and Electron Probe Analysis*, ASTM Spec. Techn. Publ., 485, American Society for Testing and Materials, Philadelphia.

Spurr, A. R., 1969, *J. Ultrastruct. Res.*, 26, 31-43.

Spurr, A. R., 1972, *Bot. Gaz.*, 133-270.

Stelzer, R., Läuchli, A., and Kramer, D., 1973, In preparation.

Steveninck, R. F. M. Van, and Chenoweth, A. R. F., 1972, *Aust. J. Biol. Sci.*, 25, 499-516.

Steveninck, R. F. M. Van, Chenoweth, A. R. F., and Steveninck, M. E. Van, 1973, In Liverpool workshop on *Ion Transport in Plants*, Anderson, W. P., Ed., Academic Press, London.

Sutfin, L. V., Holtrop, M. E., and Ogilvie, R. E., 1971, *Science*, 174, 947-949.

Tandler, C. J., Libanati, C. M., and Sanchis, C. A., 1970, *J. Cell Biol.*, 45, 355-366.

Tandler, C. J., and Solari, A. J., 1969, *J. Cell Biol.*, 41, 91-108.

Weavers, B. A., 1971, *Micron*, 2, 390-404.

Winborn, W. B., Girard, C. M., and Seelig, L. L. Jr., 1972, *Cytobiologie*, 6, 131-149.

Yarom, R., and Meiri, U., 1973, *J. Histochem. Cyto-Chem.*, 21, 144-152.

Ziegler, H., and Lüttge, U., 1967, *Planta*, 74, 1-17.

AN ANALYSIS OF THE FREEZE-DRIED, PLASTIC EMBEDDED ELECTRON PROBE SPECIMEN PREPARATION

F. DUANE INGRAM, MARY JO INGRAM
AND C. ADRIAN M. HOGBEN
Department of Physiology and Biophysics
University of Iowa, Iowa City, Iowa

INTRODUCTION

The high resolution capability of modern micro-probe equipment places stringent demands on the tissue preparation for intracellular electrolyte distribution studies. The results of measurements can be meaningless and misleading unless changes in tissue morphology and chemical redistributions have been limited to smaller dimensions than the resolution of the analytical instrument. Although there is not a generally accepted method of tissue preparation, the most promising preparations for the quantitative analysis of soft biological tissue for electrolytes involve the use of freezing techniques in which aqueous solutions are not used for any step of the preparation.

Freeze-drying as a preparation for histological examination dates back to the turn of the century. Early results normally were good only at the optical microscope level, however more recently techniques have been described for obtaining morphological preservation sufficient for electron microscopy. (Hanzon and Hermodsson, 1960, and Sjöstrand and Elfin, 1964). Examining surface features with the scanning electron microscope, Porter *et al.*, (1972) demonstrated preparations employing freeze-drying that were comparable

with those prepared by critical-point drying. The
preservation of morphology is not uniformly good
throughout any given sample however, and this variability coupled with the complexity of preparation dis-
courages some investigators from using freeze-drying.
(Marovitz *et al.*, 1970). Notwithstanding the capri-
cious nature of the technique, the potential for excellent morphological preservation and the unique
promise of faithfully preserving the distributions of
soluble intracellular ions commends the freeze-dried
specimen as an electron probe sample. It then becomes
a matter of some interest to examine the important
characteristics of the freeze-dried electron probe
preparation.

TISSUE FREEZING

In addition to the tissue changes which may take
place during the excising of samples from an animal,
there are two types of problems associated with the
freezing process which may compromise the preparation
for the electron probe. Both of these involve changes
in the tissue which result from slow freezing.

A slightly slower freezing of the intracellular
material than the extracellular fluid, can cause a
water loss from the cell. This results in shrunken,
distorted cells containing abnormally high electrolyte
concentrations. Such distortions have been described
by Meryman (1966) and are clearly demonstrated for
erythrocytes by Stirling *et al.*, (1972). Although
obvious in the extreme case, small but significant
changes are most insidious and may go undetected by
the electron probe operator with a resulting undetected inaccuracy in his measurements. Such changes can
only be thwarted by using freezing techniques that
insure rapid freezing of the entire specimen, and rejection of specimens with apparent cellular distortions. As demonstrated later, the electron microscope or scanning electron microscope are necessary

120

to assess the quality of the preparation.

The second type of freezing injury which is associated with slow freezing is the formation of large intracellular ice crystals which destroy and degrade the tissue morphology. Luyet (1951) stated that the cooling rate should be on the order of several hundred degrees per second to achieve "vitrification". He found that tissue specimens no larger than 0.1 mm in one dimension were necessary to realize such rapid cooling. Similarly, Hanzon and Hermondsson (1960) observed that it was necessary that specimens not exceed about 0.2 mm in one dimension to produce high quality electron microscope preparations reliably.

In actual practice, the freezing step is often the step most seriously compromised, as the extremely small tissue thicknesses of optimum size for freezing are not practical in many cases. Tissue of organs that are easily traumatized for example, may be frozen *in situ* and removed later. It is then necessary to examine the tissue samples carefully to decide where the tissue injury is too serious to allow elemental localizations. Making this decision is sometimes no simple task, and high resolution analyses should be accompanied by scanning or transmission electron microscope studies of the tissue preservation.

Sjöstrand (1967) offers several excellent suggestions for freezing different types of specimens and discusses the relative merits of a few of the popular freezing media. There is no generally applicable method, however, for freezing the various types of tissue rapidly, and one is often faced with developing a different technique for each different type of tissue or situation. In general, it is not sufficient to drop small pieces of tissue into the freezing medium for they often float on top; so some means must be provided for actually plunging the specimen into the freezing medium. This normally may be done by mounting the tissue pieces on small strips of aluminum foil and plunging the aluminum foil strip into the freezing medium with forceps.

DRYING OF THE FROZEN TISSUE

There is no consensus among the various authors on the precise temperature at which pure ice changes from the "vitreous" state to the obvious crystalline structure, however, it appears to be in the region of -120°C to -130° C (Dowell and Rinfret, 1960; and Meryman, 1966). From the work of Luyet and Geherio (1939) in which they demonstrated an elevation in the temperature for recrystallization of a thin layer of sucrose solution, it is inferred that the recrystallization of intracellular ice will take place at a temperature somewhat higher than -120°C to -130°C. The precise temperature, if it exists, is unknown. For the best tissue preservation then, a temperature close to -120°C should presumably be used. From a practical standpoint, however, MacKenzie and Luyet (1964) have produced experimental data on pure ice which indicates that a drying temperature of -78°C would require 6 times as long as drying at -65°C, and drying at -94°C would require 218 times as long.

Thus except for special circumstances with extremely thin sections, drying temperatures between -60°C and -80°C are most often selected. Unfortunately, a systematic study has not been performed to this date that relates the length of time necessary to dry the various types and sizes of tissue samples at the different temperatures from -50°C to 120°C with the quality of preservation realized for each drying temperature. Because of the long times involved, one does not wish to dry at any lower temperature than is consistent with the required preservation of cellular detail.

Both the excellent tissue preparations reported by Sjöstrand and Elfin (1964) and the high quality

Freeze-dried Embedded Specimen

SEM preparations presented by Porter *et al.*, (1972) were dried at -80°C, a few days for the former and 24 hours for the latter. In each case subsequent warm-up to room temperature was gradual, a few days in the former case and 6 hours in the latter. Hanzon and Hermodsson (1960) determined that specimens dried at -70°C for 48 hours produced acceptable electron micro-scope samples and they recommended drying at -40°C for optical microscope preparations. It thus appears that for most purposes, drying temperatures below -80°C are unnecessary.

The larger specimen blocks that we sometimes dry, on the order of 1 mm across, are normally dry in about three weeks with a drying temperature of -65°C to -80°C. The procedure we have adopted for our prepa-rations is that described elsewhere (Ingram and Hogben, 1968; and Ingram *et al.*, 1972). The procedure was first described by Hanzon and Hermodsson (1960), except that they stressed the importance of freezing smaller pieces of tissue than we sometimes use.

PLASTIC EMBEDMENT

The embedding with plastic is possibly one of the more controversial steps we employ. We assume that just as the embedding process does not destroy or seriously degrade the tissue morphology, so does it not cause noticeable translocation of the electrolytes we seek to measure. Although no salt movement is ex-pected to take place within the liquid plastic, the movement of salts on the boundary of the advancing plastic during infiltration is a possibility. The examination of many types of tissue has never reveal-ed artifacts that we could ascribe to this effect in the intracellular spaces. Hanzon and Hermodsson (1960) have suggested that the surface tension forces of the advancing plastic might be responsible for some of the artifacts they observed with the electron mi-croscope, such changes were minimized with osmium treatment prior to embedding. It is important that such changes, if they exist, be maintained at a level below the resolution capabilities of the electron probe. Although we do not normally see evidence of embedding artifact, some of the artifacts we attribute to the freezing or drying steps may actually result from the plastic embedding.

SPECIMEN SHRINKAGE

It is commonly assumed that freeze-drying is not accompanied by volume changes (Meryman, 1966). However, Miyamoto and Moll (1971) determined by photographic means that lung tissue which had been freeze-dried, embedded in paraffin and then deparaffinized had shrunken to 74% of its frozen volume. This measurement did not include the volume change associated with quick freezing a sample originally at 37°C. To check for volume changes in our plastic embedded specimens, we have compared the change in specific activity of specimens loaded with ^{22}Na that have been freeze-dried and embedded in plastic with the specific activity of the fresh specimens. By comparing the counting rate per unit volume of the embedded sample to that of the fresh tissue, the gamma emitter provides a straightforward means of assessing volume change.

Drops of 20% bovine serum albumin solution containing ^{22}Na (0.5µC/ml) were freeze-dried, fixed with osmium and embedded in plastic following our normal protocol. Aliquots of the original solution were pipetted, weighed and set aside for later counting. The embedded specimens were trimmed with a microtome on all sides. Sections were viewed with a light microscope to determine that all excess plastic had been removed. Paraffin was used for support during much of the trimming.

Trimmed specimens were counted in a deep well

gamma counter along with the pipetted aliquots. The density of the fresh solution was used to deduce the volume of each of the weighed aliquots. The volume of each of the embedded specimens was obtained from its weight and density. The density of the embedded specimen was inferred by determining the density of a sucrose solution in which the block was in buoyant equilibrium. To accomplish this, each embedded sample was placed in a sucrose solution of the approximate density of the sample. The concentration of the sucrose solution was then adjusted by adding solutions of greater or lesser density until the sample would remain suspended at any position in the fluid. The solution density was then measured by using a Krogh-Keys syringe pipette calibrated to deliver one cubic centimeter. Knowing the weight and the density of the sample, the volume was specified.

A ratio was formed of the specific activity of the embedded specimen to the average of the determinations of the specific activity of the fresh albumin solution. The results are summarized in Table I. The ratio was 1.20 ± .06 as the average and standard deviation of 16 different sample blocks prepared in three different freeze-dry runs at three different drying temperatures of -40°C, -65°C and -80°C. We have not established that a difference exists attributable to the different drying temperatures. However, one set of determinations of the shrinkage was made for specimens that had not been treated with osmium vapor before embedment. Compared with the osmium treated samples from that same set there is a difference. The samples fixed with osmium shrank 14% ± 4% whereas those not treated with osmium but prepared and analyzed in parallel with the osmium fixed specimens shrank 32%. This suggests a possible additional artifact in the preparation if the tissue is not stabilized with osmium before embedment. Hanzon and Hermodsson (1960) described specimen changes and damage that resulted from embedding specimens that had not been stabilized with osmium. As a result of these

TABLE I.

Determination of shrinkage of freeze-dried,
plastic embedded 20% albumin specimens[a]

Sample	Weight 10^{-3} g	Density g/cm^3	Volume 10^{-3} cm^3	Net ^{22}Na Counts per sec	Specific Activity cnts/sec/cm^3	Ratio Embedded/Fresh
Fresh Albumin (Ave., n=6)	100	1.054	94.9	521.88	5499 ± 51[b]	
Embedded Albumin (Dried at -80°C)	8.60	1.245	6.91	46.64	6750	1.227
	5.75	1.258	4.57	31.23	6834	1.234
	4.25	1.247	3.41	23.48	6886	1.252
	26.30	1.240	21.21	145.47	6859	1.247
	7.54	1.246	6.05	43.66	7217	1.312
	8.65	1.240	6.98	48.47	6948	1.263
Fresh Albumin (Ave., n=5)	96	1.054	91.1	581.69	6385 ± 74[b]	
Embedded Albumin (Dried at -65°C)	10.10	1.240	8.14	56.55	6947	1.088
	4.20	1.250	3.36	25.76	7667	1.201
	3.47	1.244	2.79	20.63	7394	1.158
	2.93	1.245	2.35	17.28	7353	1.152
	3.63	1.241	2.93	20.93	7143	1.119

(continued)

(Table I continued)

Embedded Albumin (Sans Os)	1.50	1.176	1.28	10.78	8448	1.323[c]
	4.10	1.175	3.49	29.08	8335	1.305[c]
	1.70	1.168	1.46	12.41	8529	1.336[c]
Fresh Albumin (Ave., n=5)			98.7	271.17	2747 ± 12[b]	
Embedded Albumin (Dried at -40°C)	5.90	1.265	4.66	15.29	3281	1.194
	8.50	1.272	6.68	22.48	3365	1.225
	5.57	1.270	4.23	13.93	3293	1.199
	3.70	1.274	2.90	9.66	3331	1.213
	2.67	1.268	2.11	6.70	3175	1.156
				Average of 16		$1.20 \pm .06$[b]

[a] Each set of embedded samples is presented with its corresponding fresh samples.
[b] Averages are presented ± S.D. of the mean.
[c] Samples not treated with Os are not included in final average.

TABLE II.

Determination of shrinkage of freeze-dried,
plastic embedded amphiuma blood

Sample	Weight 10^{-3}g	Density g/cm^3	Volume 10^{-3}cm^3	Net ^{22}Na Counts per sec	Specific Activity cnts/sec/cm^2	Ratio* Embedded/Fresh
Amphiuma Blood (Averages, n=5)	102	1.060	94.3	644	6829 ± 170	
Embedded Amphiuma Blood	3.25	1.242	2.617	22.62	8643	
	5.70	1.245	4.578	41.13	8984	
	3.75	1.232	3.044	25.85	8492	
	4.40	1.235	3.563	29.40	8251	
	5.85	1.251	4.676	41.42	8858	
				Average	8646 ± 291	1.27 ± .05

*Ratio ± the estimate of the S.D. obtained from the square root of the sum of the squares of the percent standard deviations.

observations, the nonfixed specimens have not been included in the average presented in Table I.

Volume changes induced in *Amphiuma* whole blood were assessed in the same manner as with albumin. The ratio of embedded to fresh was 1.27 ± .05, Table II.

An attempt was also made to measure the shrinkage in various types of animal tissue with ^{86}Rb. A mouse was sacrificed 5 hours after intraperitoneal injection of ^{86}RbCl. Samples of liver and skeletal muscle were obtained. Some samples were weighed and saved in vials and others were processed by freeze-drying, fixing with osmium, and embedding in plastic. Although the results indicated a shrinkage, the scatter was too great to make any definite statement about the magnitude of the volume change. For mouse muscle we found 1.23 ± .21 and for mouse liver 1.36 ± .69 as the ratios of specific activity of embedded to fresh tissue.

We have not attempted to isolate the source of the volume changes we observe. Although the shrinkage is most likely the result of the drying, it is not possible with the information we have to rule out an effect of the plastic embedding step. The only information we have about the effect of the embedding process is the one set in which samples not stabilized with osmium revealed greater shrinkage than those exposed to osmium vapor.

The preparation of biological samples is plagued with the potential for systematic errors introduced during the preparation procedure. It is mandatory that the occurrence of such errors be recognized when absolute concentration measurements are extrapolated back to the original wet tissue state.

STABILITY TO THE ELECTRON BEAM

An important requirement of the freeze-dried

plastic embedded specimen preparation technique is
that the specimen be able to withstand the rigors of
both the vacuum and the electron beam. Reliable quantitative measurements require a specimen that undergoes only inconsequential physical or chemical changes
during analysis. An obvious physical change does take
place in the plastic when it is first introduced to
the electron beam. What appears to be a track of bubbles is left by the electron beam in the virgin plastic. This track is not observed if the plastic is
first gently heated with a large, diffuse, electron
beam, or alternately, if the plastic is first heated
on a hot plate. After this initial heating or "stabilization" of the plastic, subsequent changes in the
plastic are of a sufficiently diffuse and general nature that no backscattered electron image degradation
is apparent. In our laboratory we routinely bombard
the specimen with about a 200 µm diameter, 100 nA beam
for two minutes to "stabilize" the plastic before
data gathering.

The normal mode we use for gathering data is the
generation of line scan profiles using a constantly
moving electron beam across usually a 10 µm to 100 µm
sweep. An "on-line" computer sorts out the x-ray intensities as a function of position of the electron
beam (Ingram, 1969). The examples described later
have all been obtained by sweeping ten times at 40
seconds per sweep.

To check for changes in x-ray production as a
function of time we have allowed an 8 µm by 10 µm area
sweep of the electron beam to remain in one location
on a portion of a specimen that has been stabilized
with a diffuse beam. This has been done for both a
thick block and for a one micrometre thick section
mounted on quartz. After scanning two hundred times
with a 10 kV, 50 nA electron beam at 40 seconds per
scan there was no change in the Na and Cl x-ray counting rates on the block of plastic embedded albumin.
A similar study with the one micrometre section

mounted on quartz revealed a three percent loss in x-ray intensity for Na, a six percent loss in x-ray intensity for Cl, and five percent increase in counting intensity for K after the same amount of time. This bombarding time corresponds to twenty times the amount of time the electron beam is scanned across the same amount of sample area for normal data gathering. Consequently, a scanning 10 kV, 50 nA electron beam can be used without changing counting intensities and be compatible with the obtaining of suitable counting statistics. For this purpose a computer or similar data averaging device is invaluable (Ingram, 1969).

SCANNING ELECTRON MICROSCOPE STUDIES

As routine freeze-dried samples have adequately preserved tissue structures at the light microscope level, the electron microscope must be used for a critical study of the quality of tissue preservation. The quality of preservation at the electron microscope level is often quite variable, requiring a considerable effort with serial sections to present an adequate picture of a specimen. The scanning electron microscope however, with its unique depth of field and wide latitude of magnification can include as much information of the type we require in one image as the transmission electron microscope can in a whole series. Because it presents information in such a highly visual form, the scanning electron microscope has been used to examine some of our typical plastic embedded specimens to demonstrate the type of morphological preservation one might expect. This has not been a systematic or complete study, but simply a brief look at some typical specimens.

Samples were presented for the scanning electron microscope by fracturing the plastic block to expose tissue surfaces and etching away the epoxy with sodium methoxide (1 or 2% sodium hydroxide dissolved in methanol). The etching takes place very slowly, and a constant flow of fresh solution must be provided to remove solid reaction products. Erlandsen *et al.*, (1973) have demonstrated that little detectable tissue alteration is induced by this procedure. The figures presented here are the result of etching for six hours. Fracturing the plastic block does not provide the same type of tissue surface as fracturing the frozen specimen because fracture planes do not follow the tissue surfaces as faithfully after embedding in plastic.

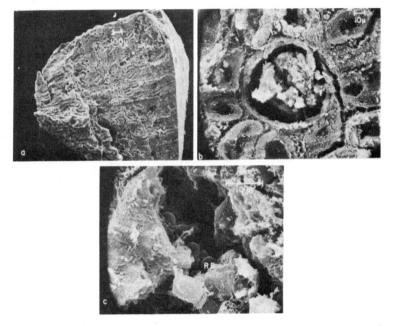

Figure 1. *A series of increasingly higher magnification scanning electron microscope photographs of a mouse kidney from which the plastic has been etched with sodium methoxide. a) low magnification image of the entire specimen featuring the etched surface; b) a higher magnification of a glomerulus; and c) a view down into the crevice apparent in (b). Note the excellent preservation of the red blood cells (RB).*

Freeze-dried Embedded Specimen

Figure 1 is a series of three scanning electron micrographs of an embedded mouse kidney prepared by freezing a whole kidney *in situ*. Figure 1-a is a low magnification image of the entire block. Figure 1-b is an area containing a glomerulus showing considerable surface debris and illustrating the lattice work structure of the cytoplasm attributed to ice crystal injury. This structure appears as voids with the transmission electron microscope. Figure 1-c is an examination at higher magnification of the fissure in the center of the glomerulus that has exposed red blood cells. The preservation of these red blood cells is excellent, revealing no signs of distortion or collapse. From the preservation of the erythrocytes we deduce that the rapid freezing was indeed rapid such that the ice crystal damage observed nearby must have come during the subsequent drying or handling.

Figure 2 is a scanning electron micrograph of a region near another glomerulus on the same block. The extent of the ice crystal injury is much more visible in this figure. The brush border, though present, contains less detail than we have observed in some transmission electron micrographs of a different specimen.

Figure 3 is a closer look along the side of an opening in a glomerulus exposing the podocytes. The processes are preserved, however they appear somewhat shrunken and withdrawn when compared with mouse kidney glomeruli that have been critical-point dried. This is most likely a freezing injury caused by the water leaving the processes before intracellular freezing. These processes are sufficiently smaller than the resolution of our normal electron probe work that these changes would likely go undetected.

Although the present examples reveal a preservation that is adequate for analysis with the 0.7 µm diameter, 10 kV electron beam we normally use for electron probe analysis, the removal of plastic for

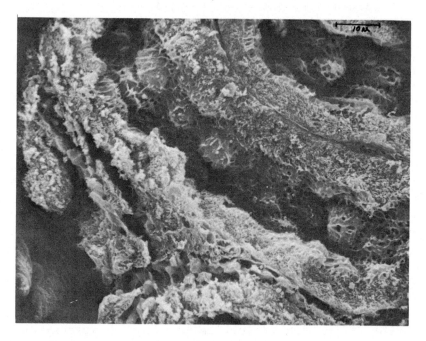

Figure 2. *A close look at the tubules next to a glomerulus just visible on the left. The lattice-like appearance of the cells is attributed to ice crystal formations, although the embedding or plastic removal processes may be responsible for some of the damage.*

tissue preservation studies is not recommended on a routine basis. The possibility of additional tissue changes induced by the plastic removal, coupled with the probable change of electrolyte concentrations, limits the usefulness of the method. Conventional electron microscope studies of thin sections are more suitable for routine studies of the quality of tissue preservation.

ELECTRON PROBE MEASUREMENTS

A few examples of different types of tissue are presented here to illustrate the nature of the

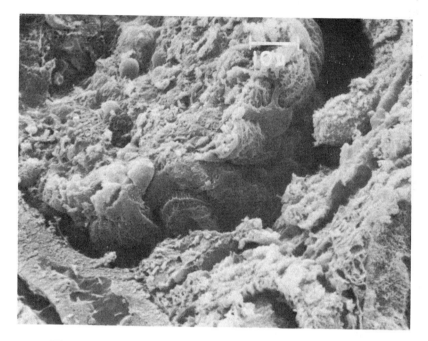

Figure 3. *An SEM presentation of a glomerulus featuring the podocytes in the center and a few red blood cells in the upper left. Surface debris obscures considerable detail on top, however an undisturbed view is obtained by looking down into features such as this glomerulus.*

information the electron probe can supply for tissues treated according to our standard protocol.

Either a face of the embedded specimen that has been cut with a microtome or a section opaque to the electron beam is coated with about a 20 nm layer of carbon and presented to the electron beam. The desired location for analysis is determined by reference to the light microscope and to either the backscattered electron or sample current images on the oscilloscope. Data is gathered by continuously scanning the electron beam across the sample. A LINC-8

135

computer (Digital Equipment Corp.) is used to sort out the x-ray counting intensities from the three detectors and to monitor the backscattered electron signal. Line scan profiles are built up by storing the computer data in arrays that represent the electron beam position. (Ingram, 1969).

X-ray intensities are converted into numbers representing the concentrations of K, Cl and Na at each point by referencing the counting rates to pure crystals of KCl and NaCl as secondary standards. These crystals have been previously "calibrated" by comparing the x-ray intensities from the pure crystals with the x-ray intensities from prepared freeze-dried, plastic embedded albumin standards blocks containing known concentrations of K, Cl and Na. (Ingram *et al.*, 1973). The uncertainties in the calibration coefficients are 3.9%, 4.7% and 3.9% for K, Cl and Na respectively. Using the criterion for absolute detection limits that the signal must be three times the square root of the off-peak background to be unambiguously detected, we find 4 mEq/kg for each of K and Cl and 6 mEq/kg for Na as the absolute detection limits.

The calibrations have been performed with a 10 kV, 50 nA electron beam. The calibration would have to be repeated if it were desired to operate at a different accelerating voltage. Modest differences between the physical characteristics of the embedded unknown and the embedded albumin standards, such as the volume changes discussed earlier, may exist and any treatment of the reliability of the electron probe measurements must recognize this possibility.

MOUSE KIDNEY PREPARATION

As mouse kidney tubules collapse immediately upon removal of a kidney from the animal, an entire kidney was frozen *in situ* to obtain electron probe specimens.

Freeze-dried Embedded Specimen

An exposed kidney of an anesthetized mouse was sur-
rounded with a modified plastic cup. The intact kidney was doused with chilled liquid propane followed
by liquid nitrogen. The kidney was maintained under
liquid nitrogen while it was removed from the animal
and shattered into small pieces. Small chips of the
outer portion of the cortex, on the order of 1 mm
across, were transferred to the freeze-dry apparatus
and prepared according to our standard protocol. Figures 1, 2, 3 are scanning electron microscope presen-
tations from one of these specimens. Figure 4 is a
representative set of electron line scan profiles of
a portion of another block from the same set. The line

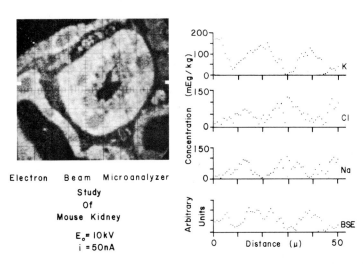

Electron Beam Microanalyzer
Study
Of
Mouse Kidney
E_o= 10kV
i =50nA

Figure 4. *Line scan profiles of the K, Cl and
Na distributions across a proximal mouse kidney tubule
depicted on the left. The picture is the oscillograph
image of the electron backscatter signal (BSE). Ten
repeated 40 second sweeps with the electron beam have
been used to generate the line scan profiles. The tick
marks on either side of the photograph delineate the
location of each end of the line scan.*

137

scan images were obtained by making multiple exposures of the computer oscilloscope images. The lettering was applied later. Featured with the line scan profiles is a backscattered electron oscillograph image of the region scanned by the electron beam to produce the line scan profiles. The profiles in Figure 4 were generated by sweeping the electron beam from the nucleus of the distal tubule on the left through the center of the proximal tubule presented in the oscillograph image. As we typically operate the electron probe, it is capable of distinguishing the differences between the cytoplasm and the nucleus and between the cytoplasm and the brush border. The detailed structure of the brush border is below the resolution of the 10 kV, 50 nA electron beam with our electron probe. Structures of smaller size, such as the mitochondria or centrioles could be seen only by using thin sections and a smaller electron beam diameter. Such components would not be well preserved in all tubular cells, and if it were desired to study such structures a considerable effort might be required to find sections of the specimen with the fine detail sufficiently well preserved for accurate analysis.

FROG RED BLOOD CELL

The red blood cell poses one of the more difficult cells for electron probe study when prepared by freeze-drying. The cell is capable of rapid equilibration with changing osmolality of the bathing medium, so any delay in the intracellular freezing is accompanied by rapid water loss with consequent increase in intracellular ion concentrations. The flattened structure of an already small cell requires great care in analysis to ensure that the electron beam samples only intracellular material and does not penetrate through and out of the cell on the opposite side. To guard against this possibility we cut serial sections, using the center 3 μm section for electron probe analysis with a 10 kV electron beam and the two 1 μm sections from either side for study with a light microscope. Each cell examined with the electron probe is determined to be present in both of the outer adjacent sections to ensure that the section the electron probe analyzes comes from the interior of the cell. This procedure was followed in measurements presented in Figure 5 of the K, Cl and Na distributions across a frog red blood cell.

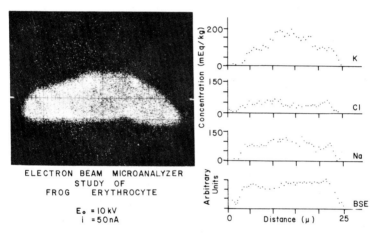

Figure 5. *Line scan profiles of the K, Cl and Na distributions across a frog red blood cell. Ten repeated 40 second sweeps have been made between the tick marks on either side of the osillograph representation of the backscattered electron signal on the left.*

The following analysis documents the precision to be expected in analyzing a biological specimen subsequent to the problems inherent in the preparation, the reliability of the absolute values being compromised by preparation artifact. The averages of the concentrations in the cytoplasm of the frog erythrocyte with the estimate of the standard deviations obtained by

139

folding in the estimates of the standard deviations of the calibration coefficients with the uncertainty in the integrated x-ray intensities were: 92.8 ± 4.4 mEq/kg; 39.3 ± 2.4 mEq/kg; and 70.5 ± 3.8 mEq/kg for K, Cl and Na respectively. Similarly, the concentrations in the nucleus were found to be 172.6 ± 7.5 mEq/kg; 36.0 ± 2.3 mEq/kg; and 94.9 ± 4.8 mEq/kg for K, Cl and Na respectively. The x-ray intensities were corrected for off-peak background and referenced to pure KCl and NaCl crystals as explained earlier.

This serves to demonstrate the precision we place on a single measurement of intracellular electrolyte concentrations. With the 10 kV, 50 nA electron beam in our 1963 ARL EMX (Serial No. 59, modified), the resolution we expect is limited by an 0.7 μm electron beam diameter and about 2.5 μm electron penetration. It is not practical then, when operating in this mode, to attempt studies on cells much smaller than these frog red blood cells.

The estimates of the standard deviations that have been presented give an indication of the precision of the measurements, but should not be construed as statements about the accuracy with which these numbers can be extrapolated back and used to represent the concentrations in the original wet tissue. It is because of the unavailability of a means of determining the accuracy of this extrapolation, while impressive high precision measurements are quoted, that the recognition of artifact and sample alteration takes on such importance.

Although precautions were taken against sampling the outer surface as well as the cytoplasm in the frog erythrocyte measurements presented here, the Na concentrations appear higher than one might expect. Attempts in our laboratory to correlate electron probe measurements of *Amphiuma* red blood cells with the results of wet chemistry have not been encouraging. We attribute the variable, high intracellular concentrations we find to slow freezing alterations in which water is withdrawn from the cell before freezing as discussed earlier. This can not explain the high Na measurements presented for this frog erythrocyte as the K and Cl signals would also be high if water loss were to take place during freezing.

FROG SKIN EPITHELIUM

Figure 6 depicts line scan profiles obtained from frog skin epithelium. Abdominal skin of *Rana catesbiana* was mounted in a flux chamber with normal Ringers (119 mM Na^+, 110 mM Cl^-, and 2.5 mM K^+) bathing each side. After the spontaneous potential was allowed to stabilize for one hour, new solution was added and it was short-circuited and allowed to stabilize for one-half hour. The chamber was drained, taken down, and a small piece of skin was excised from the center of the preparation and rapidly frozen in chilled liquid propane. It was estimated that about ten seconds elapsed between interrupting the imposed flux chamber conditions and immersing in chilled propane. After freezing, the skin specimen was prepared by the standard freeze-dry techniques.

The line scan profiles begin on the left outside the skin and pass across the epithelium into the corium on the right. The backscattered electron signal is not as useful in frog epithelium as in some tissues so that the base of the epithelium is more clearly depicted by the K signal than the backscattered electron signal. The outer cell layer of the stratum granulosum, the first reactive cell layer, was observed to be swollen in the fashion described by Voûte and Ussing (1970) for the short-circuited frog skin. The outermost cell in Figure 6 is conspicuous, having a low K and high Cl and Na concentrations. The K concentration is one fifth that of the rest of the epithelium. This low K concentration in the first cell layer is that expected with no change taking place in

141

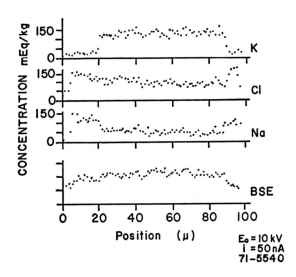

Figure 6. *Line scan profiles of the K, Cl and Na distributions across a short-circuited frog skin epithelium that had been mounted in a flux chamber with normal Ringer's solution bathing both sides of the skin. The short-circuited preparation was produced by biasing the normal spontaneous potential to zero volts by means of an external voltage source.*

the total K content of the swollen cell. Assuming a spherical shape, the cell would have to swell but 1.7 times its normal diameter to dilute the K by the amount observed. Light microscope examination reveals the cells in this outer layer to be swollen on the order of twice normal size.

The Na concentration of the swollen cell plus or minus the estimate of the standard deviation was 116 ± 7 mEq/kg. Within the uncertainties, this is the same as the concentration of the Ringer's solution bathing the skin. Unfortunately, the electron probe can not distinguish between bound Na and free Na, and without further information it is unknown if the intracellular Na concentration is indeed the same as the Ringer's solution.

One should be wary of attempting studies of this sort demanding a precision of better than 5 to 10% as it would require a great effort to realize such confidence limits with the techniques we are illustrating. This results from the uncertainties in our calibration coefficients which are 3.9%, 4.7% and 3.9% for K, Cl and Na respectively. However, if one is only interested in ratios rather than absolute concentrations, higher precision can be realized.

The frog skin data is illustrative of a class of experiments of extremely interesting studies for which the electron probe is ideally suited. Experiments such as this on isolated epithelia do not tax the resolution capabilities of the electron probe.

CONCLUSIONS

The sample preparation still remains the most critical limiting factor in the high resolution localization of electrolytes in soft biological tissue. Although elaborate, fickle, expensive and time consuming, the freeze-dry technique can produce specimens that are sufficiently stable under the electron beam to allow for analytical studies with a precision on the order of 5 to 10%. The interpretation of these values in terms of their original concentrations prior to preparation remains the crucial step however, and the errors involved in this extrapolation may well overshadow the precision of the measurements as determined by the standards and the counting statistics.

ACKNOWLEDGEMENTS

The authors would like to thank Mr. William Priester for his technical assistance and Dr. Gene Shih for his assistance with the SEM.

F. D. Ingram et al.

REFERENCES

Dowell, L. G., and Rinfret, A. P., 1960, *Nature*, <u>188</u>, 1144-1148.

Erlandsen, S. L., Thomas, A., Wendelschafer, G., 1973, *Proceedings of the Workshop on Scanning Electron Microscopy in Pathology*, Chicago, 349-356.

Hanzon, V., and Hermodsson, L. H., 1960, *J. Ultrastructure Research*, <u>4</u>, 332-348.

Ingram, F. D., 1969, *Proceedings of the Fourth National Conference on Electron Probe Analysis*, Pasadena, Calif., p. 27.

Ingram, M. J., and Hogben, C. A. M., 1968, in *Developments in Applied Spectroscopy*, W. K. Baer, A. J. Perkins, E. L. Grove, Vol. 6, Plenum Press, New York, pp. 43-64.

Ingram, F. D., Ingram, M. J., and Hogben, C. A. M., 1973, in press, *Proceedings of the Eighth National Conference on Electron Probe Analysis*, New Orleans, La, pp. 62A-62E.

Ingram, F. D., Ingram, M. J., and Hogben, C. A. M., 1972, *J. Histochem. Cytochem.*, <u>20</u>, 716-722.

Luyet, B. J., 1951, in *Freezing and Drying*, R. J. C. Harris, Ed., Institute of Biology, London, p. 77.

MacKenzie, A. P., and Luyet, B. J., 1964, *Biodynamics*, <u>9</u>, 177-191.

Marovitz, W. F., Arenberg, I. K., and Thalmann, R., 1970, *Laryngoscope*, 80, 1680-1700.

Meryman, H. T., 1966, in *Cryobiology*, H. T. Meryman, Ed., Academic Press, New York, 1-114.

Freeze-dried Embedded Specimen

Miyamoto, Y., and Moll, W., 1971, *Respiration Physiol.*, 12, 141-156.

Porter, K. R., Kelley, D., and Andrews, P. R., 1972, *Proceedings Fifth Annual Sterescan Symposium*, Kent Cambridge Scientific Co., Chicago, 1-19.

Sjöstrand, F. S., 1967, In *Electron Microscopy of Cells and Tissues. Instrumentation and Techniques.* Academic Press, New York, 1, 188-221.

Sjöstrand, F. S., and Elfin, L. G., 1964, *J. Ultrastructure Research*, 10, 263-292.

Stirling, C. E., Schnieder, A. J., Wong, M., and Kinter, W. B., 1972, *J. Clin. Invest.*, 51, 438-451.

Voûte, C. L., and Ussing, H. H., 1970, *Exp. Cell Res.*, 61, 133-140.

DISCUSSION

QUESTION: How is background determined?

ANSWER: Background is taken as the average of off-peak counts obtained from a location about five peak-widths above the characteristic peak location and an equal distance below the peak. Care is taken that an element such as osmium that is known or likely to be present in the sample does not have a high order line at the location where background is measured. The use of pulse height analyzers simplifies this task.

QUESTION: Do you find a variation in the potassium signal with the thickness of the carbon layer over the specimen?

ANSWER: Fresh crystals of KCl and NaCl are mounted on the sample block next to the specimen. The same,

approximately 20 nm carbon layer that covers the
specimen also covers the crystals. These crystals
serve as secondary standards for quantitation and
hence compensate for any variability that may result
from differences in coating thickness from one sample
to the next. We have examined pure crystals that
were placed at positions in the evaporator such that
one set received four times as thick a carbon coating
as the other set. The set of crystals with the thicker coat of carbon had x-ray intensities reduced by 3%
for each of K, Cl and Na. A similar set of measurements was made on crystals covered with approximately
twice the thickness of carbon we normally use compared with a set of crystals containing eight times
the normal thickness. This time, increasing the
thickness of carbon by a factor of four resulted in
decreases in x-ray intensity of about 7% for K, and
8% for Cl and Na. Thus the counting on pure crystals
is reasonably insensitive to coating thickness when a
thin layer of carbon is used, and elaborate precautions to insure uniform coatings are not necessary.

MEASUREMENT OF MASS LOSS IN BIOLOGICAL SPECIMENS UNDER AN ELECTRON MICROBEAM

T. A. HALL AND B. L. GUPTA
Cavendish and Biological Microprobe Laboratories, Cambridge, England

During the microprobe analysis of biological tissue sections, the electron beam removes material locally from the irradiated microvolumes. Because this effect stands in the way of the fully quantitative analyses which are required in certain studies, we have set up an on-going programme to measure the loss of mass, and to learn how to prevent it or how to cope with it. A detailed description is to appear elsewhere (Hall and Gupta, in press). Our brief report here is meant to convey the salient results to date and to encourage others to use the same technique.

The beam-induced loss of mass seems to be of two types. We will not be concerned here with the slow type (Höhling *et al.*, in press) which is due presumably to rise in temperature and is controllable by the application of heavy evaporated coatings and by the use of low beam currents. The second type, the subject of this paper, occurs very rapidly (too rapidly for detection in routine X-ray microanalysis) and is thought to be due to the escape of volatile products after the beam-induced rupture of chemical bonds (Stenn and Bahr, 1970).

One problem in following the rapid loss as it occurs is to have a sufficiently sensitive means of detection. In the work reported here we have used the continuum X-ray signal, recorded over a wide energy band (3 — 21 or 4 — 16 keV), as a measure of local mass under a static microbeam. With a Si(Li) detector placed 30 or 40 mm from the specimen (Figure 1) feeding into a multichannel analyser operating in the multichannel scaling mode, with counting intervals of

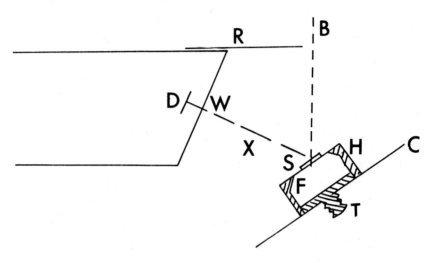

Figure 1. *Experimental arrangement.* B, *electron beam;* X, *X-rays;* S, *specimen;* D, *detector;* W, *beryllium detector window;* H, *aluminium holder;* F, *celloidin film;* R, *radiation shield;* T, *stub;* C, *cold stage. (Figures 1-5 are from Hall and Gupta, J. Microscopy, in press).*

0.9 sec, we have found that the method is sufficiently sensitive for many studies. Sensitivity could be improved by moving the detector closer to the specimen but this necessitates countermeasures against extraneous background more stringent than we have yet instituted, and in any case the sensitivity of the method is probably insufficient for studies with section thicknesses less than 1 μm or beam diameters less than 1 μm.

The X-ray intensities to be expected in a given situation can be estimated from an equation derived

Mass Loss Under an Electron Beam

from a formula of Dyson (1956):

$$N(V_1, V_2) \; \delta E = 2.8 \times 10^{-6} \; Z \; (ln \; V_2/V_1) \; \delta E$$

where $N(V_1, V_2) \; \delta E$ is the average number of continuum quanta generated within the quantum energy band from V_1 to V_2 as a single electron dissipates energy δE within the specimen, V and E both being in units of keV, and Z is the average atomic number of the specimen. In thin sections of soft tissue the average total energy loss per electron δE is approximately proportional to local mass per unit area, and for a 30-keV beam, δE is approximately 0.8 keV for a section of 100 $\mu g/cm^2$.

A further problem of the continuum method is the presence of background radiation which is not generated in the specimen, and it is essential to arrange the geometry so that this background is only a small part of the recorded signal.

A typical observation with an unembedded freeze-dried frozen section is shown in Figure 2. The tissue is from the salivary gland of a blowfly larva. This and subsequent graphs show X-ray continuum intensity vs. time, which we take to be the curve of irradiated mass vs. time. Approximately 30% of the local mass is lost within a dose of 4×10^{-10} Coulomb/μm^2, in rough agreement with the data of Bahr *et al.* (1965).

Evaporated Al coatings, 40 nm on the support film and 40 nm on the upper face of the tissue section, did not reduce this rapid loss, although when higher beam currents are used, such coatings are known to reduce the later loss due presumably to rise in temperature (Höhling *et al.*, in press).

The loss of mass in a section of Araldite is more gradual (Figure 3). Our observations on Araldite-embedded tissue sections have been variable, ranging from the pattern of Figure 2 to that of Figure 3.

149

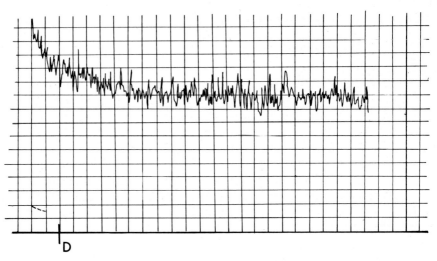

Figure 2. *X-ray continuum intensity vs. time. Section cut frozen and dried. Unfixed, unembedded section, cutting thickness 4 μm. Single, 1-nm carbon coat. 30 kV, 1 nA, 10-μm spot. Scale 15 sec/cm. Stage not cooled. The dashed line indicates the contribution of the thin plastic supporting film. The vertical marker "D" in Figures 2 — 6 corresponds to a dose, measured approximately, of 4×10^{-10} Coulomb/μm^2.*

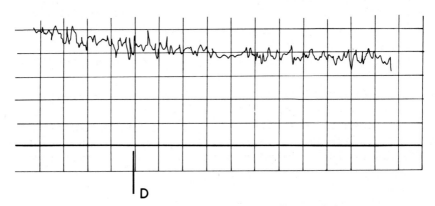

Figure 3. *Section of Araldite alone. Cutting thickness 0.7 μm. Single, 2-nm carbon coating. 30 kV, 1 nA, 10-μm spot. Scale 8 sec/cm. Stage not cooled.*

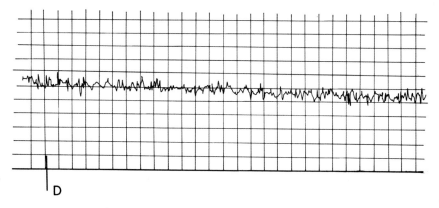

D

Figure 4. *Section cut frozen from fixed block, and dried. Irradiated on cold stage. Glutaraldehyde fixed, unembedded section, cutting thickness 4 μm. Single, 2-nm carbon coating. 30 kV, 1 nA, 10-μm spot. Scale 15 sec/cm. Stage temperature -180°C.*

With unembedded dried frozen sections the loss of mass is strikingly retarded by lowering the specimen temperature. Figure 4 is from a run with the specimen stage held at -180°C., and Figure 5 shows successive runs on adjacent microareas of a section with the stage at different temperatures.

We have just begun to work with frozen specimens which are *not* dried or sublimated. Figure 6 shows an encouraging run with such a specimen on a cold stage. It is our only successful run of this type to date; others have yielded curves like Figure 2 or even curves indicating mass increasing with time due to condensation or other contamination; but we hope with improved technique to reproduce Figure 6 routinely. With the cold stage we expect either to reduce the loss of mass to a negligible amount, or else at least to retard the effect to such an extent that the initial mass can be well determined before significant loss occurs.

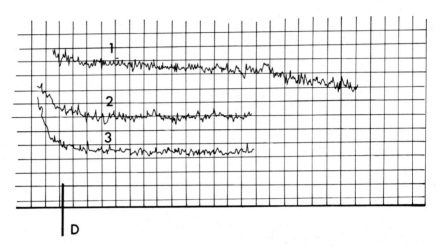

Figure 5. *Section, cut frozen and dried, irradiated at various temperatures. Unfixed, unembedded section, cutting thickness 4 μm. Double, 40-nm aluminium coating. 30 kV, 1 nA, 10-μm spot. Scale 15 sec/cm. Stage temperature changing. Run 1: From -180°C to -100°C. Run 2: From -65°C to -53°C. Run 3: +18°C.*

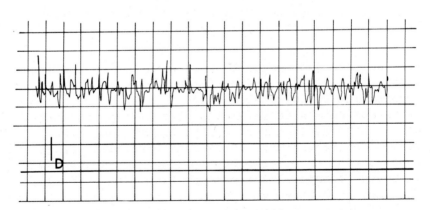

Figure 6. *Frozen specimen in hydrated state. "Iced" suspension of cells. 30 kV, 1 nA, 6-μm spot. Scale 14 sec/cm. Stage temperature -140°C.*

REFERENCES

Bahr, G. F., Johnson, F. B., and Zeitler, E., 1965, *Lab. Invest.*, 14, 1115-1133.

Dyson, N. A., 1956, Ph.D. Thesis, University of Cambridge (quoted in Green and Cosslett, 1961, *Proc. Physical Society*, 78, 1206).

Hall, T. A., and Gupta, B. L., in press, *J. Microsc.*

Höhling, H. J., Hall, T. A., Kriz, W., von Rosenstiel, A. P., Schnermann, J., and Zessack, U., in press, *Proc. Internat. Symposium on Modern Technology in Physiological Sciences*, Munich, July 1971.

Stenn, K., and Bahr, G. F., 1970, *J. Ultrastruct. Res.*, 31, 526-550.

DISCUSSION

COMMENT: I think we have to distinguish between two effects: first, radiation damage; secondly, mass-loss effect. Radiation damage is independent of temperature — there is always the same radiation damage rate. Mass-loss depends on temperature. At -150°C. or so, the rate of mass-loss may be greater than at room temperature. It is important to find out the best temperature to minimize the loss of mass.

The loss of mass can be masked by contamination, giving a constant total mass and the appearance of no loss from the specimen.

RESPONSE: Of course there is the possibility of a balance between the rate of mass-loss and the contamination rate, but our contamination rates have been low, and given the specific characteristic loss curve that we see in uncooled specimens, it seems quite unlikely that the contamination would precisely balance out this curve at low temperature.

I agree that low temperature does not prevent
radiation damage. Undoubtedly the rupture of chemical bonds still occurs, but the specimen seems to be
stabilized in some way so that the escape of material
is either slowed up or prevented.

When we use the cold stage, we do not know the
local temperature in the specimen itself. At a stage
temperature of -150°C, the specimen is at an intermediate temperature, which might be more favorable
than -150°C., as the speaker suggested.

COMMENT: There is evidence from high-voltage electron microscopy that at low temperature you don't
reduce the damage, but you only see the damage when
you bring the specimen up to higher temperature.

COMMENT: Is there a possibility that bound water es-
capes during the irradiation of tissue, and that this
may change at low temperature?

RESPONSE: Yes, this seems likely.

COMMENT: Someone has also shown losses with embedding material like Epon.

RESPONSE: Yes, we as well as others have also seen
loss effects in epoxy resins.

COMMENT: Can we take it then that you feel that reducing the temperature is worthwhile not only for your
iced specimens but also for more routine freeze-dried
and embedded specimens, and that the reduction of
mass-loss should be useful?

RESPONSE: Yes, it does seem that way.

COMMENT: Heat may enhance the curing of an epoxy
resin, inducing cross-linking and making it tougher.

Mass Loss Under an Electron Beam

The observations on epoxy resins may be the result of the combination of this effect, and the boiling out of volatile components.

COMMENT: In the first milliseconds, or the first moments of irradiation — what happens?

RESPONSE: The hypothesis given in my paper is taken from the literature, especially the review paper by Stenn and Bahr. They say that in the first moments chemical bonds are ruptured and radiation products are formed, including such products as CO_2, H_2, and N_2, which escape. As to why this escape is inhibited at low temperature, I do not know in detail, but the inhibition does seem to occur. I really have nothing further to offer in response to this question. (Additional response added at a later date: The inhibition may be partly in the step between bond rupture and formation of the volatile products. At low temperature the mobility and reactivity of the primary products of the bond rupture may be too low to lead to the formation of the volatile products.)

COMMENT: We could probably live with C, N and H going away if we knew that things like Na were not being lost. At the recent meeting in Chicago (IITRI, SEM '73), there was a report of the installation of a quadrupole mass spectrometer in a scanning electron microscope. It would be nice to use a system like that to find out what is coming off.

RESPONSE: Yes, I know that Tom Hutchinson is going to try to do that.

COMMENT: Professor Koenig has done this with a mass spectrometer in a transmission electron microscope, and he found that the first product coming out was NO_2, and then CO_2, and then H_2.

COMMENT: What about other elements? What about Na and K?

REPLY: He didn't find anything.

COMMENT: Was he analysing tissue, or what?

REPLY: He was investigating tissue, and then support films, collodion and Formvar.

COMMENT: We have studied 2-μm frozen-dried sections
with a beam of $\frac{1}{2}$ nA in a scanning electron microscope,
and have observed no loss of Na, K or Cl with time.
We see the same levels with the beam scanning over a
very large area (100 μm^2), or over a much smaller
area. The only loss observed was in the "background"
(non-characteristic X radiation), in which we saw a
loss of 10% in albumin sections.

COMMENT: In epoxy-embedded ultrathin sections, in
20-second counts for Na, we have observed a falling
off in the first few counting intervals and then a
leveling off.

COMMENT (by the author): We have to keep in mind that
in terms of current per unit area, 10 nA put into a
spot of diameter 100 nm (1000 Å) corresponds to a current of 1000 nA put into a spot of diameter 1 μm. In
earlier microprobe work we were accustomed to putting
10 nA into a 1-μm spot with no trouble, but much less
current can be tolerated in a spot of 100 nm.

COMMENT: But the early microprobe work was with bulk
specimens where all the energy is absorbed. In thin
sections much less energy is absorbed.

RESPONSE: No, I am thinking of early work with tissue
sections, where certain safe levels of current were
established, but these levels were with larger spots.
Of course the physical conditions of heat dissipation
vary with spot size. But to exemplify the point: For
the analysis of K and Na, as standards we have used
sections of the minerals sanidine and albite. We have
found that we can burn the K and Na out of these

standards very easily in the EMMA with a small spot
and low current, although the standards were stable
under much higher currents, with a larger beam, in an
older microprobe analyser.

COMMENT: Do you think that the temperature at the
site of the electron beam is really at, say, -180°C?

RESPONSE: No, of course not.

COMMENT: So, in fact, you vaporize everything, or
you break all the bonds, the temperature is so high.

RESPONSE: We break the bonds. We are not vaporizing
everything. We have seen that with a cold stage, in
fact, a hydrated specimen remains frozen. (To achieve
this you do have to take precautions to have good
thermal conductivity, so that the specimen is effectively cooled.) The specimen temperature is at some
intermediate value, and we do not know what it is.

COMMENT: One effect many times reported in the scanning microscope — we have seen it routinely — is that
if the beam scans a raster and moves slightly, you
reduce contamination and/or etching markedly, as
against leaving the beam in one spot.

RESPONSE: We have seen this effect too, in our mass-
loss curves as well, where we get a classical loss
curve and then sometimes a rise which we attribute to
specimen shift.

COMMENT: That's the big reason why we use our scanning beam, together with our computer to sort out
where it is located. We find that the specimen damage
is practically eliminated by not allowing the beam to
stop. The scanning is over lines of 25 to 100 μm, repeated many times until enough counts are accumulated.
That really makes our system go. If we had to put the
beam on one spot for, say, 40 seconds or so, the area
gets so badly distorted that you can't take a second

reading.

COMMENT: We have worked on specimens air-dried at room temperature as well as on deep-frozen cooled specimens. With red blood cells at room temperature, the characteristic peak intensities dropped under repeated analyses over the course of one hour. At liquid nitrogen temperature the peak intensities as well as the continuum were fairly stable. These were bulk specimens. Can you explain why, at room temperature, the characteristic peak intensities went <u>down</u>? One would expect, with loss of mass, an <u>increase</u> in the relative concentrations of Na and K and the characteristic peaks should go up.

In other specimens we found that the intensities of the characteristic peaks did go up, as expected from the loss of mass, but in certain specimens including the red blood cells, the effect was definitely a decrease in the peaks.

RESPONSE: In the bulk specimens, one can speculate that there may be something like the effects which have been seen in glass, where the electrolytes may be driven away to regions (below or around the irradiated volume) where the beam doesn't reach them. Or there may be overheating, and these elements may be lost even more rapidly than the organic mass, or lost to a greater extent. (Additional reponse added at a later date: The early loss of organic mass may have been too rapid to be manifest in any way in these measurements, i.e. so rapid that the first count was affected as much as the later ones. With rise in temperature there may be further change in organic mass, but this may level off fairly soon, while there may be a continuing loss of volatile elements like K or Na going on for a much longer time.)

THE PREPARATION OF BIOLOGICAL MATERIALS FOR X-RAY MICROANALYSIS

PATRICK ECHLIN AND ROGER MORETON
*Botany School and Biological Microprobe Laboratory,
Department of Zoology,
University of Cambridge, England*

INTRODUCTION

In order to carry out X-ray microanalysis of biological material, it is necessary to stabilize the
cells and tissues so they can withstand the high vacuum which obtains inside the instrument. For many
cells and tissues it is sufficient to follow a fairly
conventional regime of fixation, dehydration, embedding and sectioning, or in the case of single cells
to fix and dehydrate and examine the cell intact.
Such techniques, while providing much useful morphological information, suffer from the singular disadvantage in that they invariably result in a considerable loss of low molecular weight soluble substances,
ions and electrolytes from the tissues. This is because the fixatives used to stabilize cellular structural features also cause gross permeability changes,
and the subsequent treatment with dehydrating agents
result in a rapid egress of material from the cell.

In the past few years a number of workers have
been attempting to devise techniques and methods to
stem this unfortunate flow of materials from cells
and tissues, and nearly all investigators agree that
the cryobiological approach offers the best solution
to the problem. The essential feature of this approach is that the water in the biological material

is rapidly converted to ice and this dramatic and sudden increase in the viscosity of the cell fluids locks in position the soluble substances we wish to analyse. Before cooling, the material may or may not be carefully and lightly fixed and/or impregnated with an appropriate antifreeze agent.

The water may now be removed from the material by means of sublimation in a high vacuum or the material may be sectioned in the frozen state, and the sections freeze-dried or examined in the still frozen state. There are many variations which may be composed around this cryobiological theme.

This paper presents one of these approaches. The rationale for the experimental procedure has been dictated primarily by the type of specimen we wish to examine. Our principal interest is in the localization and distribution of low atomic weight ions and electrolytes such as Na^+, K^+, Ca^{++} in a wide range of plant and animal tissue. We have devised techniques for three different types of specimens, the bulk sample, cells or cellular fractions, and sectioned material. While much of the preparative procedure is the same for all three types of specimen, important differences do exist and these will be expounded where appropriate.

Although the basic procedures have been satisfactorily worked out, we do not consider that this present exposition is the definitive statement on the subject. What follows is, in many ways, an interim report and much further work remains to be carried out.

Parts of the preparative procedure have already been described in the papers by Echlin (1973), Echlin and Moreton (1973) and Boyde and Echlin (1973).

Preparation for X-Ray Microanalysis

METHODS AND RESULTS

Pre-treatment

We have tried wherever possible to minimize or
avoid altogether, any chemical treatment before the
rapid freezing procedure. All previous work, and our
own experience, show that any form of fixation causes
some change in the properties of the cells. Similarly, we have avoided the use of cryoprotectives and
artificial nucleating agents. Such materials, while
minimizing ice-crystal damage in the specimen, are
physiologically incompatible with the types of analyses we are attempting.

The bulk specimens are damp-dried in a stream of
warm air. This process must be carefully monitored
and is usually carried out under a dissecting microscope. Specimens are attached to the specimen holder
using a thick paste of colloidal graphite. The pre-
treatment of the cell suspensions and cell fractions
is invariably dictated by the exigencies of their particular preparative procedure. The suspensions are
spread as a monolayer on the support film using a very
fine Pasteur pipette. This is best done under a dissecting microscope, and only repeated trials and experience will show when a suitable monolayer is obtained. We find that it is best to put a large drop
on the supporting film, to let it stand there for a
minute or so, and then gently to draw off the bulk of
the drop leaving a thin layer. Once the layer is obtained it should be allowed to nearly dry before being frozen. Material to be sectioned is either attached and orientated on the microtome chuck and
quench frozen, or small pieces of the tissue are frozen and subsequently fixed to the chuck.

Supporting Films

Sections and cells suspensions are supported on
a thin aluminized nylon film. The nylon film is made

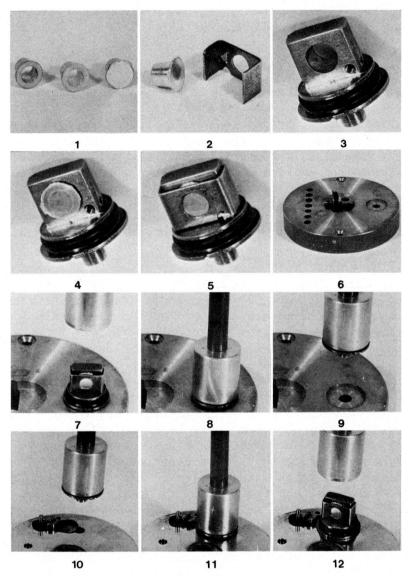

Figure 1. *Aluminium collar assembly, empty, with nylon film and with aluminized nylon film.* Figure 2. *Aluminium collar and phosphor-bronze retaining clip.* Figure 3. *Specimen stub holder.* Figure 4. *Specimen stub holder with aluminium collar in position.* Figure 5. *Specimen stub, with aluminium collar and clip in*

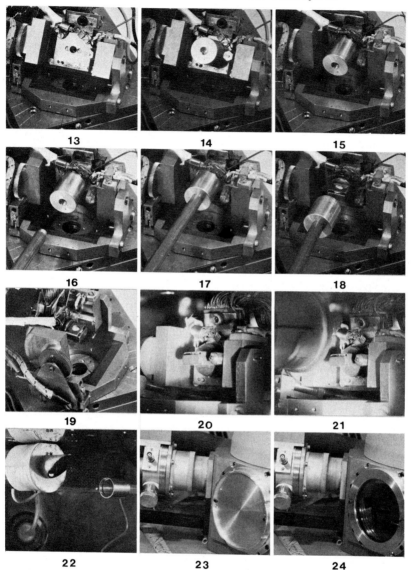

position. Figure 6. *Orientating jig for assembling specimen holder in the cryostat. and under liquid nitrogen.* Figures 7, 8, 9. *Capping and removal of specimen from cryogen or cryostat.* Figures 10, 11, 12. *Placing and decapping specimen on evaporator cold table.* Figures 13-18. *Placing specimen on microscope*

163

according to the recipe obtained from Hall (personal communication). 50 grams of nylon chips are dissolved in 500 mls of isobutyl alcohol by gently heating to about 90°C in a liquid paraffin bath. This stock solution may be kept in a suitable container and may be used later simply by re-heating to c. 75°C. Small, conical aluminium collars (Figure 1) are thoroughly cleaned, and the tops roughened with emery cloth, and placed in the holes of an aluminium plate. The plate plus collars is submerged in a shallow tray 5 cm deep by 60 cm × 30 cm filled with cold clean tap water. Care should be taken to avoid any air bubbles in the collars and between the collars and the aluminium plate. The surface of the water is cleaned by running a glass rod along the top and a thick nylon film is cast in order to cover the entire surface. This is done by gently pipetting as a continuous stream, the equivalent of 8-10 drops of the dissolved nylon mixture, which has been allowed to cool to about 60°C. This film is allowed to solidify and is then removed from the surface with the glass rod. Such treatment ensures a thoroughly clean and dust free water surface. The nylon film may now be cast as follows. The equivalent of 4 to 5 drops is placed on the water surface as a continuous stream. This quickly spreads but does not cover the entire surface. When viewed by reflected light a variety of iridescent colours may be seen, and the submerged plate with collars is gently manoeuvred so as to be immediately under a continuous region of purple or gold coloured film. The water level is now dropped and the nylon film lowered onto

cold stage and removing the cap. Figure 19. *Specimen in correct orientation for X-ray microanalysis and scanning transmission microscopy.* Figures 20, 21. *Relative position of specimen and X-ray detector as seen inside the microscope.* Figure 22. *Relative positions of X-ray detector, secondary electron detector and cap removal device.* Figures 23, 24. *Energy dispersive detector and chamber window.*

the plate and collars. This should be done rather slowly and it is advisable to cover the entire surface of the water bath during the procedure. It is important that the nylon film is tightly stretched over the surface of the collars to avoid any wrinkles in the film. Once the film is attached and partially dry, the plate and collars may be lifted from the water bath and the excess film cut away. The plate and collars are carefully dried and stored over night in a desiccator. The nylon film is aluminized by placing the entire plate and collars in a vacuum evaporator and evaporating between 15 and 20 nm of aluminium on the surface. Full details of the apparatus and procedure are given in the papers by Echlin and Hyde (1972) and Echlin and Moreton (1973).

The individual aluminium-nylon coated collars may now be gently removed from the plate as they are required. The collars may be re-used simply by removing the excess film and thoroughly sonicating first in isobutyl alcohol and then in acetone.

Rapid-freezing

The rationale underlying, and the procedure for, quench-freezing bulk samples and tissues for subsequent sectioning have been discussed elsewhere (Echlin and Moreton 1973, Boyde and Echlin 1973) and will not be dealt with further. However the procedures for cell suspensions and cell fractions are sufficiently different to warrant inclusion in this present paper.

The cell suspension is placed on the aluminized support film as previously described, and the collar placed in the conical hole of the copper specimen stub (Figures 3, 4). The aluminium collar is held in position with a small phospher-bronze spring clip (Figures 2, 5). The copper stub is based on the design of the stub used in the Cambridge Stereoscan S4 microscope and essentially holds the specimen at right angles to the normal viewing position.

The stub, collar and clip assembly is then rapidly plunged into a small bath of liquid nitrogen, which is in turn placed within another bath of liquid nitrogen. The double bath system ensures that the temperature of the inner bath is lower than the boiling point of nitrogen and prevents any bubbling which easily obscures some of the later manipulative procedures. The liquid nitrogen baths are kept covered to prevent water condensation. Liquid nitrogen is not a good cryogen because of the closeness of the melting and boiling points of this gas. However for cell suspensions and cell fractions, which maximally are between 5-10 μm, the theoretical cooling rates of objects of this size are sufficiently fast to cause only minimal ice crystal damage. We have tried using fluorocarbon as a cryogen and while it is possible to achieve better cooling rates, the excess of the fluorocarbon which inevitably gets transferred to the vacuum evaporator, can create some difficulties in obtaining a sufficiently good vacuum pressure to enable coating to be carried out. Once frozen, the stub assembly should remain below the surface of the liquid nitrogen to prevent ice contamination.

Sectioning

Sections between 1 and 5 μm are cut at approximately -80°C on a Slee microtome kept in a large capacity cryostat. The interior of the cryostat is kept dry and at a slightly positive pressure using cold dry nitrogen. The sections are cut dry using a steel knife and are transferred to the aluminized nylon films by the following procedure. The aluminium collar covered with the aluminized nylon film is gently pressed onto the sections lying on the back of the steel knife. It is important that the collar assembly is slightly warmer than the section. The supporting collar and film is then inserted into the conical hole on the specimen stub (Figures 3, 4) and clipped into position using the phosphor-bronze spring (Figures 2, 5). The stub assembly is firmly held in an orientating jig (Figure 6) while these transfers are being made. All these procedures are carried out at -80°C inside the cryostat and great care must be taken to ensure that ice does not condense onto the specimen.

Specimen Transfer

It is necessary to transfer the specimens from the freezing and sectioning equipment to the coating unit and eventually to the scanning microscope. These transfers must be done in such a way as to prevent ice condensing onto the specimen and thus obscuring important detail. We have made small caps to fit over the entire specimen while it is under the cryogen and which seal onto the stub by means of an O ring at the base (Figure 7). The capping procedure has already been described (Echlin and Moreton 1973) and need not be repeated here. The capped specimen may now be transferred to the cold table of the vacuum evaporator (Figures 8, 9, 10, 11 and 12).

Specimen Coating Procedures

The coating procedures for bulk specimens have already been described (Echlin and Moreton, 1973) and we will confine our comments to the procedures adopted for cell suspensions and fractions, and thin sections. The capped stub holder is transferred to the cooled cold table of the evaporator. Once the capped stub is orientated so that the section or the suspension faces the evaporation source (Figure 11) it is locked into position by gently turning the small screw. The evaporation chamber is now pumped down to its working vacuum and the cap removed from the specimen by means of an insulated rod which passes through a rotary vacuum seal and screws into the cap (Figure 12). A thin layer (c. 10-15 nm) of aluminium or carbon may now be evaporated onto the surface of the specimen. It is, however, most important that the pressure should reach at least 10^{-3} N m^{-2} before any evaporation is attempted, otherwise the aluminium is deposited as a thin

167

black film. Once the coating is completed the cap is
replaced over the specimen and the whole unit allowed
to come up to atmospheric pressure using cold dry nitrogen. The capped specimen may now be removed from
the vacuum evaporator and transferred to the cold
stage of the microscope. In some instances we have
avoided the second coating layer, but its presence
certainly does add to the stability of the specimen.

Specimen Examination

The specimens are examined and analysed at low
temperatures using the Cambridge Hot and Cold Module
fitted to the series 200 stage. We have made a number
of changes to the design of both the Hot-Cold module
and the stage, and the details are given in the paper
by Echlin and Moreton 1973. Essentially what we have
done is to increase the working distance to 20 mm
making it possible to insert the "nose" of an energy
dispersive X-ray detector between the specimen and the
final lens, so we can carry out X-ray analysis at very
close range with the specimen horizontal (Figures 21,
22). The specimen bucket can be tilted up to 135°
about a horizontal axis, and rotated through about
100°. The capped specimen is locked into position on
the cold stage (Figures 13, 14), cooled to its lowest
temperature (about -175°C) and the microscope pumped
down to its working vacuum. Small identifying marks
on the stub ensure that the stub is inserted in the
correct configuration with the sections or suspensions
horizontal to the beam axis. During pumping down, the
cold stage is turned and rotated so that the capped
stub faces the small rear right-hand port on the
microscope (Figure 15). The cap may now be removed
using a small rod passing through a vacuum seal, (Figures 16, 17, 18) and the specimen turned through 90°
so that it faces the energy dispersive detector (Figures 19, 20) which is inserted through the small rear
left-hand port of the microscope chamber. The X-ray
detector is on a retractable mounting and once the
specimen is in the correct position the nose of the

detector is brought forward to within a few mm of the specimen (Figure 21). The de-capping procedures and positioning of the X-ray detector are facilitated by the presence of a window fitted to the large left-hand port of the microscope chamber (Figures 23, 24). The inside of the chamber is lit with small lights and the whole area may also be completely blacked out.

X-ray microanalysis may be carried out on a variety of specimens. Morphological visualization of the specimen is carried out by means of either reflected secondary electrons or by transmitted electrons. A hole has been made in the base of the cold module assembly and this extends through the stage assembly (Figures 16, 17, 18). This will enable us to fit a solid-state transmitted-electron detector which we are confident will provide us with satisfactory images.

DISCUSSION

Although we are fairly certain that it will not be possible to fix material before quench freezing and still retain the ions and electrolytes we wish to analyse, we have insufficient evidence from our own work to make a catagorical statement to this effect. In this connection it would be worthwhile undertaking a methodical analysis of the leaching effects of various fixation and tissue stabilization procedures, to see if some compromise might be reached between the deleterious disruption in permeability of the specimen and the marked improvement in preservation which can be achieved by using these materials. A similar investigation should also be undertaken with regard to the use of cryoprotectives and artificial nucleators. Yet in spite of the absence of fixation, and cryoprotectives, quite a high degree of morphological preservation may be obtained. Our results show that in thick (c. 0.5–1.0 μm) frozen-dried sections prepared according to our method we can resolve (morphological detail) between 100 and 200 nm which is better than the

169

resolution which can be obtained by X-ray microanalysis. Once we have the transmitted electron detector in operation we are confident that we may resolve morphological detail of between 40 and 50 nm. We have recently been able to resolve about 100 nm in 2-μm thick sections cut from material which had been fixed with glutaraldehyde, dehydrated and embedded in Araldite. The sections were unstained and encapsulated between two thin layers of aluminium. We have calculated that glutaraldehyde-fixed, dehydrated and Araldite-embedded material has a mass per unit area equivalent to a frozen (i.e. retaining water as ice) section of similar thickness cut from an unfixed frozen specimen. This material has made a suitable test object for some of our early experiments. Although problems of contrast still remain, particularly in unstained specimens, the scanning transmission mode of specimen signal acquisition does give better signal to noise ratios and does allow the image to be electronically enhanced. Although methods have been devised to enable X-ray microanalysis to be carried out on bulk material, this approach is not the method of choice. This is due to the spreading effect of the electron beam in the specimen which gives rise to much poorer spatial resolution.

The capping procedures we have devised successfully prevent ice forming on the specimen. This is most important, for not only does ice obscure the specimen, it also increases the scattering of the electron beam. The coating procedures, although rather tedious to perform, are vital to the success of our experimental approach. By encapsulating the frozen specimen between two layers of conducting material we may keep it cold and prevent it warming up. The metal used for coating must obviously not interfere with the analysis, and we are currently investigating the use of other substances.

It is also important to minimize the opportunities for specimen contamination during the preparative procedures. A cold object in a vacuum system is always the focus for any contamination. It is, for example, going to be necessary to improve the quality of the vacuum in the specimen chamber, for with present instruments the contamination rate in the absence of anti-contaminating devices is probably of the order of between 10 and 100 molecular layers a second!

ACKNOWLEDGEMENTS

The authors are grateful to the Science Research Council for their continued interest and financial support for this project. This paper represents part of a study on the examination and analysis of frozen biological material being carried out in collaboration with Dr. T. A. Hall, Dr. B. Gupta and Professor T. Weis-Fogh. Thanks are also due to Mr. N. Cooper for his valuable assistance with cryomicrotomy and Mr. D. Tyler for designing and constantly modifying all the pieces of instrumentation associated with this experimental procedure.

REFERENCES

Boyde, A., and Echlin, P., 1973, Freezing and freeze-drying — a preparative technique for scanning electron microscopy. In *Scanning Electron Microscopy*, O. Johari, Ed. IIT Research Institute, Chicago, Ill. pp. 759-766.

Echlin, P., and Hyde, P. J. W., 1972, The rationale and mode of application of thin film to non-conducting materials. In *Scanning Electron Microscopy*, O. Johari, Ed. IIT Research Institute, Chicago, Ill. pp. 137-146.

Echlin, P., 1973, Scanning microscopy at low temperatures. *Journal de Microscopie*, in press.

Echlin, P., and Moreton, R. B., 1973, The preparation, coating and examination of frozen biological materials in the scanning electron microscope. In *Scanning Electron Microscopy*, O. Johari, Ed. IIT Research Institute, Chicago, Ill. pp. 325-332.

DISCUSSION

LÄUCHLI: Suggested using the freeze-substitution exchange as a means of preserving biological tissues for X-ray microanalysis. This technique uses an organic substance such as acetone in which the ions are not soluble to replace the frozen cellular fluids following quench freezing.

ECHLIN: While this technique would be applicable to situations where the analysis is to be carried out on bound ions, it is likely that freeze-substitution will result in the migration of unbound ions due to the removal of the water from the cells. A careful comparative study should be made between frozen dried, frozen hydrated, freeze-substituted and conventionally fixed and embedded tissue to ascertain the degree of loss of ions and electrolytes.

MUIR: Suggested a way to remove the small pieces of ice and frozen cellular debris which can contaminate the specimen following freeze fracture. She considered it would be practical to remove the debris by ultrasonic cleaning in chilled fluorocarbon.

ECHLIN: This would be impracticable because it would mean "washing" the sample in a number of changes of fluorocarbon, and each change would increase the chances of further atmospheric contamination.

KAUFMANN: Expressed surprise that we could observe no significant ice contamination on our specimens following metal evaporation onto frozen fractured material maintained at -150°C at 5×10^{-7} torr.

172

Preparation for X-Ray Microanalysis

ECHLIN: In spite of the use of a cold trap in the
evaporator, it is probable that there is some ice
contamination due to condensation of residual water
in the vacuum chamber. This contamination has not,
however, been found to interfere with the morphological appearance of bulk specimens, and we do not have
sufficient data to know whether it seriously affects
frozen cell suspensions and sections. We are careful
to keep the inside of the evaporator as clean as possible, and always back-fill with dry nitrogen. Our
attempts at evaporating aluminium onto specimens have
so far resulted only in a thin black (oxide?) deposit.
It would seem that freshly evaporated aluminium is
particularly reactive and can readily form oxides even
with ice maintained at low temperatures as well as
with traces of water vapour in the vacuum. We are now
attempting to clean up the vacuum even further. A
more serious problem is the amount of ice contamination we are getting inside the microscope column in
spite of dry nitrogen back-filling and the presence
of a large area of cold surfaces. Evidence has already been presented showing an <u>increase</u> in the mass
of specimens and we are now making a concerted effort
to improve the quality of the vacuum in the microscope
and to construct a small ante-chamber to the microscope through which the specimen may be inserted onto
the cold stage.

FUCHS: Commenting on the desirability of being able
to etch frozen fractured surfaces showed that minimal
etching of such surfaces occurred even when a filament heated to 2000°C was held a few cm away from the
surface.

ECHLIN: This is quite surprising, but may be a consequence of there being no cold trap in the system on
which the sublimated water may be re-deposited. We
have had some success with ice removal using electron
beam etching using high beam intensities.

HALL: In answer to a question on the geometry of the

X-ray detector in our system, stated that we employ a standard SEM with a working distance of about 20 mm from lens to specimen, and that the X-ray detector looks down on the specimen with a takeoff angle of about 30°.

LECHENE: Commented on ice crystal formation in biological specimens.

ECHLIN: The aim of quench freezing is to remove the heat from small samples as fast as possible in order to minimize ice crystal formation. Unfortunately, this heat removal is rapidly slowed down by the ever increasing ice layer which forms on the specimen. Ice has a very low thermal conductivity and is a principal contributory factor to ice crystal damage. Vitrous or amorphous ice is only formed when water vapour condenses onto very cold surfaces and is unlikely to exist in biological systems. Once amorphous ice is raised above the re-crystallization point (c. 130°C for pure water) it goes into a regular crystalline form, the crystals of which increase in size with an increase in temperature. Any work with frozen hydrated specimens should be carried out below -130°C or as near to this point as is practicable. In the absence of cryoprotectives and artificial nucleators, all we can hope to do is get ice crystals small enough not to seriously disrupt the specimens we are examining. Theoretically the faster we can remove heat from the specimen, the smaller are the resultant ice crystals. There is still some disagreement over the practical cooling rates which we can obtain, but it is somewhere between 2000 and 5000°C/sec.

Note added in proof (December 1973). Our most recent preparative procedures for x-ray microanalysis of frozen-hydrated sections are described in *Nature*, 247 1974 (in press).

ENERGY-DISPERSIVE X-RAY MICROANALYSIS IN SOFT BIOLOGICAL TISSUES: RELEVANCE AND REPRODUCIBILITY OF THE RESULTS AS DEPENDING ON SPECIMEN PREPARATION (AIR DRYING, CRYOFIXATION, COOL-STAGE TECHNIQUES)

J. GULLASCH* AND R. KAUFMANN**
*Siemens. A.G., Karlsruhe,
**Dept. of Clinical Physiology,
University of Düsseldorf

INTRODUCTION

In the past decade, many techniques have been employed to analyze chemical composition in cellular and subcellular structures in order to get more information about their biological function. The electron probe microanalysis became a very promising tool in this work. However, to get relevant results new preparation techniques have to be developed, especially when the distribution of intracellular electrolytes should be analyzed. Fixation and preparation techniques suitable for classical transmission electron microscopy or for scanning electron microscopy were primarily developed to preserve the structures of the biological material. However, for the analysis of water-soluble inorganic compounds, these techniques are not suitable at all since many of the steps employed necessarily induce heavy redistributions, particularly of the small inorganic electrolytes. Therefore, during the last three years new preparation techniques preventing such intolerable redistribution of water soluble inorganic compounds, as for instance sodium, magnesium, potassium, calcium, and chlorine have been proposed, discussed and at least in some parts, technically established.

First attempts were made by Echlin (1971), and by Kaufmann and Gullasch (1972) using hydrated specimens which were shock-frozen and investigated in a deep frozen state. Preliminary results obtained by both groups were promising enough to continue that work further. Since in many other laboratories it was felt that much simpler techniques such as air drying, freeze-drying or freeze-substitution might be sufficient to prevent ion-redistribution, a comparative study has been undertaken in order to get more information about the benefits of various techniques in one respect, namely: Prevention of ion-redistribution during preparation and during electron microprobe analysis. Two rather simple biological specimens were used:

a) human red blood cells, because their intracellular ion composition is fairly well known (and presumably rather uniform) and

b) *Tetrahymena pyriformis*, since the results obtained could be related to Coleman's (1972) data that analyzed *Tetrahymena* with a wave-length-dispersive electron probe microanalyzer.

METHODS

The preparation procedures employed in our experiments were:

a) air drying and investigation at room temperature,

b) air drying and investigation at LN_2 temperature,

c) shock-freezing in boiling nitrogen and investigation of the hydrated specimen at LN_2 temperature, and

d) shock-freezing in melting nitrogen and investigation of the hydrated specimen at LN_2 temperature.

The instrument used was the ETEC-autoscan with a LN_2

cooling stage, and a specimen transfer system which prevented surface ice formation during specimen transfer. The specimen support was a T-shaped carbon carrier which could be fitted into a copper block in close contact with the cold copper. The stage could be heated up by an internal heater in order to bring the specimen surface into a temperature range of about -120 to -90°C. At these temperatures the sublimation pressure of ice comes close to the ambient pressure of about 10^{-5} torr. Sublimation starts rapidly and can be observed in the SEM-picture. Simultaneously details of the intracellular structures become visible. Thus it was possible to direct the electron beam on the area of interest selected for microanalysis. The freeze-etching could be stopped by turning off the heater. The depth of etching could be fairly well controlled by observing the SEM-picture. We found an etching depth of a few 100 Å to be sufficient for surface observation. On the other hand this depth is small as related to the depth of X-ray excitation in a bulk specimen.

Specimen transfer could be done through an airlock-system by means of a special container, which allowed the specimen to be carried under liquid nitrogen, thereby avoiding any ice formation on the specimen's surface.

The energy-dispersive X-ray detector from Kevex was chosen with special concern to the problems involved in energy-dispersive X-ray analysis of biological specimens. The movable detector (150 eV resolution at MnKα, window 7.5 μ, area 12 mm^2) could be placed as close as about 1 cm to the specimen surface. The take off angle was further optimized by an inclination of the detector towards the specimen plane. Special considerations were given in selecting the proper energy range of primary electrons. We found that the best voltage in this work seems to be 10 kV for several reasons:
      a) Specimens have been investigated

uncoated. That means, high electron energy may create electrostatic charging on the specimen surface.

b) The energy of the interesting X-ray lines ranges between 1 keV and 3.7 keV (Ca). For this energy range, the best signal to background ratio was found with a primary electron energy of 10 keV.

c) Scattering radiation from the areas surrounding the specimen (Cu at the cold stage, Fe at the pole piece of the final lense) can be neglected only at 10 kV or below.

d) The power dissipation in the specimen which increases the specimen surface temperature should be kept as low as possible.

For the sake of a rough quantitative analysis in our hydrated specimens, we had first to perform measurements on standards. The standards used were droplets of shock-frozen electrolyte solutions containing known concentrations of sodium, potassium, chlorine and phosphor. After transfer into the SEM-cold stage, we obtained the spectra shown in Figure 1. The detection limit under the indicated conditions ranges below 3 mM. By increasing the measurement time we can expect a much smaller statistical error and hence the detection limit will probably be below 1 nM. The detection limit for sodium is about twice that of the other compounds since the 7.5 μ Be-window absorbs about 50% of the sodium Kα-radiation. The figure further demonstrates a fairly linear relationship between peak-height and ion-concentration within the range of measured concentrations.

RESULTS

The first series of measurements on biological objects were performed on human red blood cells obtained in the usual way; by smearing a small droplet of fresh blood on the polished surface of our carbon specimen support. After the smear had been air-dried, the uncoated red blood cells were analyzed in the

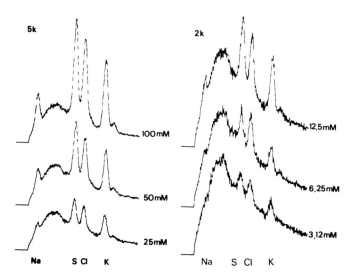

Figure 1. *Energy-dispersive X-ray spectra ob-*
tained from deep frozen standard electrolyte solution
of different ionic concentrations (10 kV, 10^{-10} A,
100 sec).

instrument at room temperature.

Figure 2 illustrates the spectra obtained from
one single red blood cell if electron microbeam irradiation is continued over a period of about 1 hour.
At first glance it is evident that — with the exception of Si — all peak intensities of the characteristic radiation decrease during the prolonged measuring
period, whereas the brems-continuum is virtually unchanged. From these results it must be concluded that
in biological specimens investigated at room temperature not only a loss of organic material during electon probe analysis must occur (as described by Hall
(1973) and others), but there must be also changes in
the element concentrations as well. Since in the
present example, the specimen is considered to be
thick as related to the depth of excitation, a loss
of organic material is not expected to reflect in a
decrease of the contiuum radiation. However, such a

179

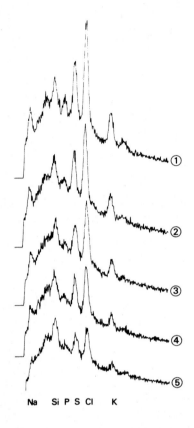

Figure 2. *Energy-dispersive X-ray spectra of a human red blood cell, air dried, investigated at room temperature (10 kV, 10^{-10}A, 100 sec).*

loss of organic matter should result in a relative increase of inorganic element concentration and hence the P/B ratio of the characteristic radiation should increase. Surprisingly enough, in the case of red blood cells the opposite is true, i.e. a significant decrease of the characteristic radiation is observed. It will be shown later that in other biological specimens, as for instance in air dried *Tetrahymena pyriformis* investigated at room temperature, the apparent loss of mass in fact led to the expected increase of element concentrations. These controversial findings are difficult to interpret. Whatever might be the cause of this effect it can be prevented by cooling the specimen down to LN₂ temperature. This is

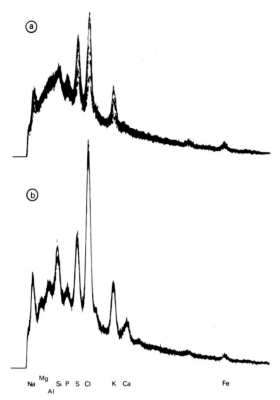

Figure 3. *Energy-dispersive X-ray spectra of a human red blood cell, (a) air dried; repetitive analysis at room temperature (10 kV, 10^{-10} A, 100 sec) and (b) air dried; repetitive analysis at LN_2 temperature (10 kV, 10^{-10} A, 100 sec).*

demonstrated in Figure 3. At the upper part six spectra of red blood cell sequentially registered at room temperature are superimposed and compared with a family of spectra obtained from a different air dried blood cell cooled down to LN_2. The picture clearly demonstrates the "stabilizing" influence of cooling on the intensity of characteristic radiation during prolonged measuring periods. In this finding, we agree completely with the results obtained by Hall (1973).

Let us now consider the intracellular ion composition as revealed by the characteristic X-ray spectra obtained in air dried erythrocytes. If we look at the present spectra we must admit that the peak intensities measured correspond by no means to the concentration pattern of electrolytes to be found in healthy human red blood cells. Particularly, the sodium and chlorine content is much too high, whereas the potassium concentration appears to be extremely low.

Better results could be obtained if small droplets of native blood were shock-frozen in LN_2, freeze-fractured and analyzed at LN_2 temperature. The spectra shown in Figure 4 were all collected from one single hydrated erythrocyte kept at LN_2 temperature during a prolonged measuring period of 1 hour. The

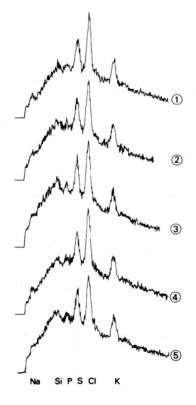

Figure 4. *Energy-dispersive X-ray spectra of a human red blood cell, hydrated, cooled at LN_2 temperature (10 kV, 10^{-10} A, 200 sec).*

spectra shown are fairly reproducible, and the intensities of the characteristic radiation are much more
consistent with the expected concentration of intracellular electrolytes.   There is clearly less sodium
and chlorine and much more potassium present in the
hydrated deep-frozen erythrocytes when compared with
the air-dried specimen previously shown.   To make this
finding statistically more convincing, populations of
red blood cells both air-dried and hydrated were analyzed at LN_2 temperature.   Figure 5 shows six spectra
randomly selected from the first group.   The second
group is presented in Figure 6.   It can be easily

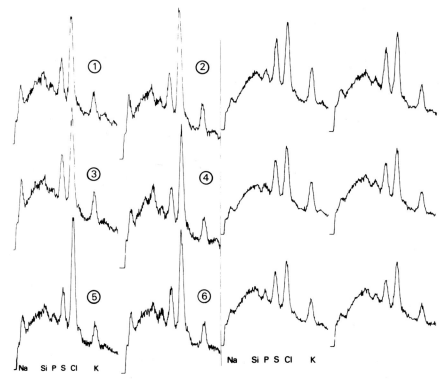

Figure 5. *Energy-dispersive X-ray spectra of 6
human red blood cells randomly selected, air-dried,
analyzed at LN_2 temperature (10 kV, 10^{-10} A, 100 sec).*
   Figure 6. *As 5, but hydrated, instead of air-
dried.*

recognized that in both groups, in shock-frozen hydrated specimen as well as in the air-dried preparations, the spectra show rather small variations. However, from the viewpoint of the expected intensity pattern, the results obtained from hydrated cells approach the physiological situation much better than the spectra of air-dried specimens.

Nevertheless, if the peak intensities measured in hydrated shock-frozen preparations are related to the corresponding spectra of our deep-frozen water standards, there is still a discrepancy between the electrolyte concentrations measured and those to be expected in a living red blood cell. This concerns particularly the concentration of potassium which from the peak intensities measured amounts to about only ½ of the expected concentration, and the concentration of chlorine which exceeds the usual chlorine content of erythrocytes by a factor of about 2. Therefore, we considered the possibility that the cooling speed in boiling nitrogen might be too slow to prevent ionic redistribution during the ice formation. Consequently, the cooling rate has been considerably increased by using melting nitrogen as a cooling medium, thereby preventing the Leydenfrost phenomenon.

If red blood cells are shock-frozen in melting instead of in boiling nitrogen, the characteristic X-ray intensities are very close to the expected values. This is demonstrated in Figure 7 where the results obtained with the various preparation and investigation procedures are comparatively summarized.

Now I would like to proceed to our measurements done in *Tetrahymena pyriformis*. Cells were grown at 16°C in the usual medium containing 2% proteose peptone. For analysis, cells were taken from cultures in their stationary growth phase. Prior to investigation, aliquots of 5 ml medium were centrifuged at 1200 rpm for 7 min. The supernatant was discarded and the pellet resuspended in about 0.25 ml of fresh cultural

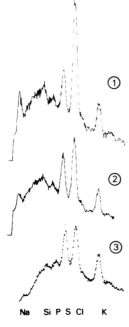

Figure 7. *Energy-dispersive X-ray spectra obtained from a human red blood cell air-dried and analyzed at LN$_2$ temperature (1); hydrated, frozen in boiling N$_2$ and analyzed at LN$_2$ temperature (2); hydrated, frozen in melting N$_2$ and analyzed at LN$_2$ temperature (3).*

medium. Droplets of that medium containing about 50 cells/mm^3 were placed on our carbon support and either shock-frozen in LN$_2$ or smeared into a film and air-dried. Shock-frozen specimen were freeze fractured before transferred into the instrument.

Figure 8 shows six spectra registered from one single air-dried cell at different times during microbeam irradiation continued over a period of one hour. It is obvious that the characteristic X-ray intensities, particularly those of Na, P, and K, steadily increase during continued analysis, indicating that the relative concentrations of these elements increase within the irradiated area. This effect is probably due to the decomposition of organic material induced by the electron bombardment, and the resulting loss of mass we already talked about.

Again, specimen cooling to LN$_2$ temperature prevented this effect entirely as can be seen in Figure 9.

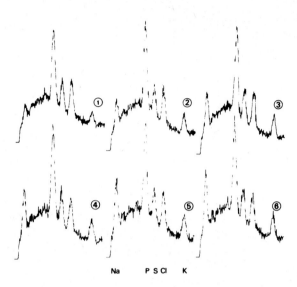

Figure 8. *Energy-dispersive X-ray spectra obtained from Tetrahymena pyriformis, air-dried; repetitive analysis at room temperature (10 kV, 10^{-10} A, 40 sec).*

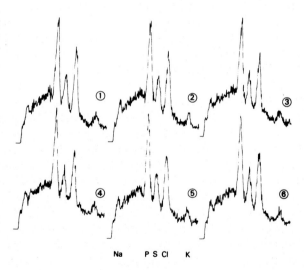

Figure 9. *Energy-dispersive X-ray spectra obtained from Tetrahymena pyriformis, air-dried, repetitive analysis at LN_2 temperature (10 kV, 10^{-10} A, 40 sec).*

It shows six spectra from one single air-dried cell obtained at different times during microprobe analysis continued over a period of one hour. If the specimen was kept at LN_2 temperature, the intensities of all characteristic X-rays measured did not change systematically over the period of observation.

Since *Tetrahymena* represents a highly organized cell, inorganic compounds are expected to be characteristically distributed over the cellular body. How this distribution pattern looks, or how it can be externally influenced, seems to be rather unknown. Here, electron probe microanalysis could supply us with a good deal of new information, provided the preparation techniques employed carefully avoid any cellular or subcellular redistribution of the compounds to be analyzed.

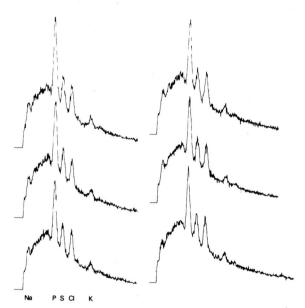

Na   P S Cl   K

Figure 10. *Energy-dispersive X-ray spectra obtained from Tetrahymena pyriformis, air-dried; analyzed at room temperature, various areas of interest (10 kV, 10^{-10} A, 40 sec).*

Figure 10 shows six spectra obtained from various areas of interest in one air-dried *Tetrahymena* analyzed at room temperature. These spectra look rather uniform indicating that either inorganic ions are evenly distributed over the whole cell, more probably, have been redistributed during specimen preparation. The only major variations found under these conditions were in the content of Ca and P.

If air-dried *Tetrahymena* were cooled to LN$_2$ temperature during analysis, the situation did not markedly change as can be seen in Figure 11. Again, the only characteristic variations of peak intensities as depending on the area of interest were found with Ca and P. In most cases where a recognizable Ca-peak was present, the P-peak was also high as compared with the spectra showing no Ca-peak.

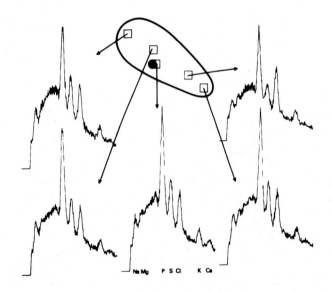

Na Mg   P S Cl   K Ca

Figure 11. *Energy-dispersive X-ray spectra obtained from different areas of interest of one single Tetrahymena pyriformis, air-dried; cooled at LN$_2$ temperature (10 kV, 10^{-10} A, 200 sec).*

Air Drying, Cryofixation, Cold Stage

This image drastically changes if, instead of an air-dried shock-frozen specimen, a hydrated specimen is analyzed at LN_2 temperature. Now, large variations of the ion content depending on the subcellular area of interest could be found.

The spectra shown in Figure 12 exhibit such gross intensity variations in all peaks of inorganic compounds with the only exception of S. However, since S is a basic constituent of the protein material it is indeed likely to be rather uniformally distributed over the whole cell body. But, in contrast to the findings in air-dried preparations, there are now variations of Cl, P, and Na over more than an order of magnitude.

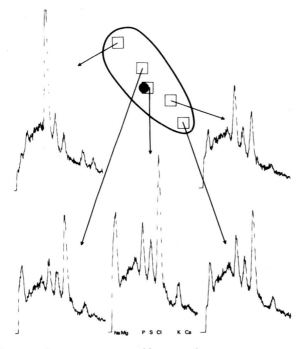

Figure 12. *Energy-dispersive X-ray spectra obtained from different areas of interest of one single Tetrahymena pyriformis, hydrated; cooled at LN_2 temperature (10 kV, 10^{-10} A, 200 sec).*

CONCLUSION

We are aware that an extensive statistical analysis is needed to make these results more reliable than they are in the present casual form. However, they appear clearcut enough to justify the following summarizing statements:

Air drying is not a suitable preparation technique if water-soluble inorganic compounds are to be analyzed by electron probe microanalysis on a cellular or subcellular level.

Electron probe microanalysis of uncoated air-dried specimen, performed at room temperature, induces further errors due to unpredictable changes of the relative element concentration during microbeam irradiation.

Shock-frozen hydrated specimens analyzed at LN_2 temperature show element distributions fairly consistent with the expected physiological concentration pattern. However, some results indicate that the cooling rate should be as fast as possible, in order to avoid ionic redistributions during ice formation.

REFERENCES

Coleman, J. R., Nilson, J. R., Warner, R. R., and Batt, P., 1972, *Exp. Cell. Res.*, *74*, 207-219.

Echlin, P., 1971, *4th Annual SEM-Symposium IIRTI*, Chicago.

Hall, T. A., 1973. This volume.

Kaufmann, R., and Gullasch, J., 1972, Beiträge zur elektronenmikroskop. Direktabbildg. von Oberflächen (BEDO) *5*, in press.

ELEMENTAL MICROANALYSIS OF FROZEN BIOLOGICAL THIN SECTIONS BY SCANNING ELECTRON MICROSCOPY AND ENERGY SELECTIVE X-RAY ANALYSIS

T. E. HUTCHINSON, M. BACANER*
J. BROADHURST** and J. LILLEY**
Department of Chemical Engineering
**Department of Physiology*
***Department of Physics*
University of Minnesota, Minneapolis, Minn.

It is highly desirable to be able to follow the process of ion motion during the performance of certain cell functions such as muscle contraction, nerve firing, and perhaps liver secretion, to name a few. Some efforts have been made in the past using highly sophisticated electron-optical tools, such as the one-million-volt electron microscope, to directly observe the microstructure of cells. These efforts have met with only limited success. Our approach has taken a slightly different form, in that we have elected to use a drastically slowed down process of molecular activity in order to obtain a picture of ion position within cells at a particular point in time. The fundamental technique which we employ is that of very rapid freezing of the tissue which effectively stops the motion of the ions at the particular point in time selected. This is obtained in practice by placing the tissue in a particular state of activity, and freezing the tissue while in that state, by contact with a cryogen such as Freon 12 at its freezing point, or a liquid nitrogen-cooled copper block. This rapidly reduces the temperature to below -100°C. Rapid cooling is required since the formation of ice crystals during

191

slow cooling destroys the tissue through mechanical
action. In addition, ion motion can take place during
the process of cooling which is undersirable for microanalysis.

Once the tissue is transformed to a frozen state,
thin sections roughly 2000 Å (200 nm) in thickness
are cut from the bulk materials using a cryomicrotome
operated at -90°C. The thin sections are collected
on a standard electron microscope grid and the grid is
placed in a special cylindrical specimen holder (pre-
cooled to -100°C) while still in the frozen state (see
Figure 1). The temperature is maintained at -90°C
in the cryogenic sectioning device by a stream of cold
nitrogen gas obtained from boiling liquid nitrogen.
This gas blanket also provides a shroud which prevents
frosting of the very cold materials during cutting and
mounting on the specimen holder.

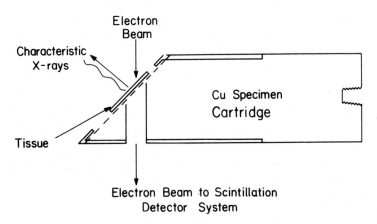

Figure 1. *Specimen holder cartridge.*

The specimen is transferred to the scanning electron microscope by means of a specially designed transfer tube or "magic wand" shown in Figure 2. The transfer tube is simply a cylindrical brass tube pre-cooled
to liquid nitrogen temperatures and fitted with a non-
conductive epoxy rod within the body of the tube to

which the specimen holder is screwed. The specimen
holder is attached to the epoxy rod and withdrawn into
the body of the tube while still within the liquid
nitrogen vapor stream. The tube itself acts as a cold
trap for wet air travelling up the tube to the vicinity of the specimen. Transfers of several minutes
can be accomplished using the tube in moist air.

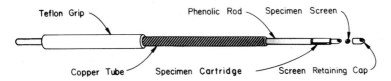

Figure 2. *Specimen transfer device.*

The specimen holder with the associated tissue
is transferred by this means to a pre-cooled (-100°C)
cold stage of the scanning electron microscope. The
cold stage is mounted on the standard S.E.M. gonio-
meter stage and in the specimen loading position is
enclosed in a dry atmosphere box shown in Figure 3.
The transfer tube is inserted through a small hole in
the side of the box and the epoxy push-rod is extended to expose the specimen to the inert dry atomosphere.
Liquid nitrogen is continuously fed into the top of
the box and a stream is directed at the cold stage of
the microscope in order to maintain a sufficiently
low temperature during transfer and to provide a cold,
dry atmosphere for frost prevention. The cold stage
of the microscope, which has been described in detail
elsewhere (Hutchinson, in press), is shown in Figure 4.
Cooling is attained via a fine stranded cooper braid
attached to a one-half inch diameter copper rod which
passes through a seal out of the vacuum and is in
contact with liquid nitrogen. Temperature control and
measurement are also attained by means of a thermistor
and resistor combination. A schematic diagram of the
system used is shown in Figure 5. The electron beam
of the scanning electron microscope enters from the
top, strikes the frozen specimen and yields transmitted
electrons which are recorded on the electron detector

Figure 3. *Dry-cold transfer box.*

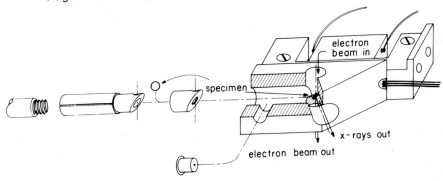

Figure 4. *Cold stage.*

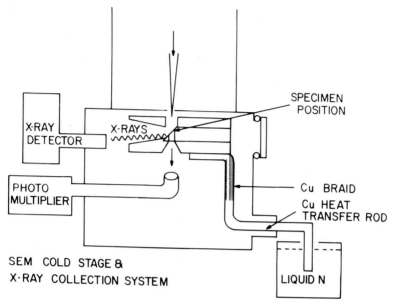

Figure 5. *Schematic diagram of the experimental arrangement.*

system situated immediately beneath the specimen. In addition, x-rays are collected on a Si-Li x-ray detector situated 3 mm from the specimen surface. In this way the 200 nm section of frozen tissue may be observed directly by electron imaging with simultaneous collection of x-rays.

The schematic of the procedure is shown in Figure 6. The electron beam of the scanning microscope is scanned across the surface of the tissue and a transmitted electron micrograph is recorded at a magnification from 20 times to 50,000 times. Structural components of the muscle tissue used in this case can be clearly discerned even in unfixed frozen tissue as seen in Figure 7. The z-lines are clearly visible in this case. The scanning beam is then put into a configuration suitable for examination of a selected area within the tissue. This takes the form of either a line scan along a z-line for instance in which the electron beam scans back and forth across the area in

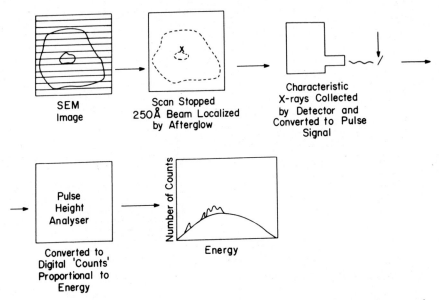

Figure 6. *Schematic diagram of the experimental procedure.*

a linear fashion or it may take the form of a spot in
which the scan beam is stopped and placed at particular positions within the tissue. In this case, although the diameter of the probe can be reduced to
roughly 20 nm, scattering within the tissue causes the
x-ray resolution to be of the order of the thickness
of the section.

X-rays are collected by the silicon (Si-Li) detector and energy-analyzed, and the characteristic
signals are stored in a pulse height analyzer. Elec-
tron-excited elements present corresponding to the
energies of the x-rays are determined from the peak
heights recorded. An x-ray analysis of the muscle
tissue on and off the z-line is shown in Figure 8.
Some changes can be seen in certain of the elements
within these spectra; however as yet the data are in-
complete as to quantitative details.

One disturbing feature of this otherwise very

196

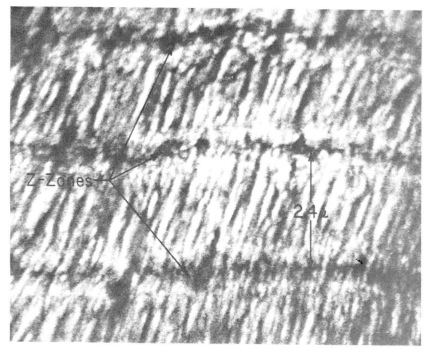

Figure 7. *High magnification scanning transmission micrograph of a 140-nm section of rabbit psoas muscle.*

powerful technique is that elements certainly are lost from electron beam-specimen interaction. This has been documented in the papers by Stenn and Bahr (1970) and in recent work by Hall (1972). Loss of elements during the time required for obtaining an adequate analysis of an area may vary between ten and fifty percent and is element selective. Lower loss figures are found in specimens at much reduced temperature.

In conclusion, we may now make elemental determinations of materials within frozen cells in a particular state of activity. As yet only muscle contraction has been studied by this technique, and that in a very preliminary way. We propose however to do other tissues such as heart and nerve tissue and to obtain time snapshots of the elemental configuration

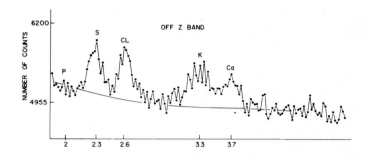

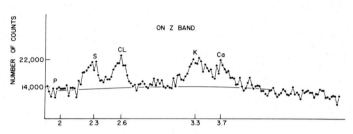

Figure 8. *Characteristic x-ray energy analysis of elements present on and off the z-line of muscle shown in the previous figure (Figure 7). The numbers along the horizontal axis give quantum energy in keV.*

within the cell during a particular stage of activity. We are also involved in efforts to determine the extent of elemental loss during electron beam specimen interaction using mass spectrometry to obtain some quantification of the curve of elemental loss versus integrated beam. We feel that the technique is immensely powerful and will most certainly provide answers to many of the very perplexing problems associated with cell function.

REFERENCES

Hall, T. A., 1972, Private communication partly reproduced in these Proceedings.

Hutchinson, T. E., Bacaner, M., Broadhurst, J., and Lilley, J., (in press), *Rev. Sci. Instr.*

Microanalysis of Frozen Sections

Stenn, K., and Bahr, G. F., 1970, *J. Ultrastruc. Res.*, 31, 526-550.

DISCUSSION

QUESTION: You say that the section is 2000 Å thick?

ANSWER: Yes, roughly. We usually set the microtome somewhere in the range 1500-2000. I don't know what I had it set for within that range on that particular day, but it was certainly not 1000 Å.

QUESTION: Did you get any mass loss during the observations?

ANSWER: Yes. This can be seen from a reduction in the characteristic x-ray signal for each element. Our laboratory is now carrying on extensive mass loss studies from frozen tissue using ultra-high vacuum and a mass spectrometer.

QUESTION: What was the accelerating voltage?

ANSWER: 15 kV. At 20 kV you start washing out detail. The resolution improves at the higher voltage but the contrast falls quickly.

QUESTION: Did you also look at the specimen by collection of backscattered electrons?

ANSWER: That was not done. It seems like a very good idea. The specimen configuration does not lend itself readily to imaging with backscattered electrons; it was set up for peak efficiency of collection for x-rays and transmitted electrons.

QUESTION: How long can you study a section?

ANSWER: As far as the morphology is concerned, for hours, literally hours, if the temperature is low

enough. The temperature of this particular section
was of the order of -100 to -110°C. We are really pressing the system to get down to that point; one can go
lower but we are not able to at the moment.

QUESTION: Is the density of the z line (in the image)
just due to more carbon, nitrogen or oxygen?

ANSWER: We simply don't know. The contrast mechanisms
are not all clear at the moment.

GENERAL DISCUSSION ON MERITS AND LIMITATIONS
OF DIFFERENT PREPARATIVE TECHNIQUES WITH
PARTICULAR RESPECT TO PRESERVATION OF
ORIGINAL ION DISTRIBUTION IN CORRELATION
TO CELLULAR STRUCTURE

The discussion opened with a consideration of the quality of the vacuum in the region of the specimen.

ECHLIN said that the pressure of 10^{-5} torr in the specimen chamber of the Cambridge S4 microscope presented some problems, as it was believed that residual water vapour could easily become deposited on frozen specimens. A way around this was to place a cold, perforated baffle above the specimen.

KAUFMANN stated that in the ETEC instrument the vacuum is of the order of 10^{-6} to 10^{-7} torr. Under these conditions they have never observed any ice formation on the specimen surface even with very long observation times. So, he thought that a better, or a cleaner vacuum could prevent ice contamination.

RUSS, speaking on the same subject, stated that the vacuum system in his microscope is not outstanding, but his group also simply surrounds the specimen with a large baffle which improves the quality of the vacuum in the specimen region. Interestingly enough, not only does the baffle decrease the water contamination (they do not get any obvious ice formation on their specimens), but it also reduces the silicon contamination quite a bit. Russ asked for anybody's comment as to where the silicon is coming from.

LÄUCHLI asked whether Russ had ever tried a substitute for the silicon in the pump oil and in other parts of the microscope.

RUSS stated that they had no silicon in the specimen area; they were not using silicon grease on the O-rings; and to the best of his knowledge there was no silicon-containing lubricant anywhere in the instrument.

LÄUCHLI asked whether contamination was still present, and RUSS replied that they still have silicon contamination. It was suggested (by an unidentified speaker) that this might be coming from the silicon detector itself, but RUSS said that the silicon detector was behind a beryllium window. However it was thought (unidentified speaker) that you could excite the detector itself by X-rays directly even behind the beryllium. RUSS replied that the observed effect was above and beyond the spurious silicon detector peak, which arises only by excitation of the silicon atoms which are not in the active volume of the detector, and which is very small. The observed effect was actually building up on the surface of the sample as a function of time. (Editorial comment added later: The spurious detector peak may be larger than suggested if it is possible for backscattered electrons to pass through the beryllium window, either because of a relatively high beam voltage, or because parts of the window may be especially thin. But the observed buildup rules out such an explanation in this case).

GULLASCH made a short comment on freeze-etching procedure. It had been mentioned that it should be possible to get good etching by heating the specimen from above. His group had tried this by putting a tungsten filament, which could be heated up to about 2200°C, very close to the specimen surface (about 5 — 10 mm away). However, they were unable to get any etching on the surface. That meant that at least in bulk specimens, by radiation alone it is very difficult to bring enough heat into the specimen to get sufficient water sublimation.

Preparative Techniques — Problems

ECHLIN said that he had tried electron beam etching of frozen specimens, and found that it was possible to remove some of the surface ice that way, but it was necessary to scan for a long time.

GULLASCH replied that what they had intended to do was not to remove ice formed by contamination, but rather to etch away the embedding ice in order to get the structural features of the "water embedded" material.

ECHLIN said that he had simply fractured his frozen samples. He had not tried to etch since he did not have a device to fracture, to etch and then to coat.

LÄUCHLI asked Russ whether he was still sticking to his assumption that in thin sections the lateral resolution is approximately equal to the section thickness.

RUSS said that they have made some measurements on Epon-embedded material, and also done a little work with frozen sections, and the two appeared to be very similar. For instance, for a 150 nm (1500 Å) section, 90% of the signal is derived from a cylinder which is about 80 nm across at the middle of the section.

LECHENE asked whether it is possible to get a vitreous state of water by the process of ice formation.

ECHLIN considered that it is impossible to get water into an uncrystallized state. Ice can be formed in two ways, as far as we know. In one way, you can come up from a low temperature, and have solid water formed on a very cold surface from water vapour in the environment. This is the nearest we come to water in a vitreous state, but electron diffraction will still show structure at about 2 nm, in other words, there are ultramicrocrystallites. Further, when the temperature rises above -130°C you get above the recrystallization point of pure ice and you go to the larger cubic ice crystals, and then to the hexagonal ice crystals. The other pathway is to go downwards from higher temperature. If we intend to cool water from, say, 20 to 25°C to well below its freezing point with the minimum amount of ice crystal formation, we want to get below the recrystallization point (-130°C) as fast as possible. Now the problem is literally heat removal, because the ice which is formed as a surface layer has a very low heat conductivity. The only way to improve the situation in the absence of cryoprotectives is to increase the freezing rate.

RUSS asked whether the best thing to do was to keep the ice crystals small enough not to show up at the resolution being used.

ECHLIN agreed, and thought that if we were working at, say, 100 nm resolution at the X-ray level, it did not matter too much if we had, say, 50 nm crystallites.

RUSS asked whether even with the formation of very small ice crystals, there was a danger of the advancing freezing front sweeping ions ahead.

ECHLIN replied that this probably would not occur if the freezing rate was fast enough, but expressed some doubt as to whether such rapid freezing had ever been achieved. Since we are talking about water in biological systems rather than pure water, nucleation would occur. If you go down from the nucleation point to the recrystallization point as fast as possible, then you get, hopefully, lots of very small ice crystals. Whether you can improve this by putting in cryoprotective agents, or by using an artificial nucleation medium is not known. He asked whether anybody had tried putting in a small amount (1%) of chloroform, which is reported to be a good artificial nucleator, to see whether there is a diminution in the ice crystal damage in biological material to which no cryoprotectives had been added.

KAUFMANN stated that during ice formation in a cell,

there is a certain period of time when the liquid and
solid phases of water are both present. During this
period the electrolyte concentration in the liquid
phase increases considerably. This implies that the
electrochemical gradient across membranes changes
drastically, which can give rise to some redistribution of ions across the membranes.

LECHENE remarked that in the process mentioned by Dr.
Kaufmann, it is thought that ice is formed as a "distilled" phase, and the electrolytes are concentrated
more and more in the remaining liquid phase of the
compartment. He asked for a comment on the fate of
the ions, in terms of the limiting case of a glass of
freezing seawater, where we should find all of the
ions at the bottom of the glass. What happens to the
ions during the freezing process when the liquid phase
is more and more concentrated?

RUSS agreed, and said that in the case of something
like seawater, it does not matter how fast you freeze
it and how small the individual ice crystals are; the
ions still get concentrated at the bottom. They may
become physically trapped between all of the ice crystals.

KAUFMANN considered the problem to be that we do not
know to what extent the biological structure can prevent this kind of separation. Perhaps the cell border,
at least, prevents ions from passing through rapidly.
Also the mitochondria, for instance, have some borders
where, we may hope, ions cannot pass through during a
very fast freezing.

SCHIMMEL made a comment relating to the paper presented by Gullasch, which included X-ray spectra obtained
from the outside of dried red cells, and spectra from
within the cells themselves. He judged it to be impossible to calibrate the spectra directly. His question was: How were the dried-cell spectra calibrated,
and by means of what standards?

General Discussion

GULLASCH replied that he had no standards other than the deeply frozen water standards described in his paper.

SCHIMMEL stated that it was then absolutely clear that you must get a different relation between the different elements.

GULLASCH agreed that this was right as far as the air-dried preparations were concerned; in that case one could only compare the heights of the different peaks.

SCHIMMEL reiterated, as his main point, that you cannot do a comparison in this way. He remarked that in one case, the water-containing specimen or standard, we have a thick specimen with the Na and K distributed mainly in a matrix of water; in the other case you have a dried cell which may have a thickness of about 2 μm or so, with the Na and K distributed mainly in carbon (i.e., in the carbonaceous matrix of the cell (Ed.)). The absorption process in the latter case must be totally different from the former, and this difference in absorption changes your relations between the different peaks.

GULLASCH said he always took the same cell for comparison, so he had about the same size of excited volume.

SCHIMMEL still took exception, saying that you cannot expect that the ratio between the Na peak and the Cl peak would be the same in both cases since each had a different matrix.

GULLASCH repeated that they did not compare different cells; they compared only the peaks obtained from one cell.

SCHIMMEL asserted that this also could not be a valid procedure, and he made his point more explicitly. In the bulk hydrated specimens the (relatively soft, non-penetrating) sodium radiation may be strongly absorbed,

and because of structural inhomogeneities, the results
may be affected by large differences in absorption
from point to point; in the case of the thin, dried
specimens you have mainly carbon, and the absorption
of the sodium radiation is much less. Therefore, in
the dried specimens, the Na signal must be increased
in comparison with the K signal.

GULLASCH said that he understood, but did not think
that absorption in the cell microstructures would have
such a large effect.

LÄUCHLI suggested that because of the different matrices found in biological systems, we should be careful
in choosing our standards when making comparisons with
air-dried cells.

KAUFMANN sought to clarify the position. He said that
it was clear that in the case of an air-dried specimen, the peak to background ratio must be better than
in the case of hydrated specimens because the mass
fractions of the analysed elements are much higher in
the air-dried specimen. But on the other hand, he did
not think that Dr. Schimmel's argument was applicable,
because the average atomic number in the air-dried matrix was not so different from the average atomic number in the hydrated specimen. So, the situation as a
whole was fairly consistent for both kinds of specimen. The only thing which they (Gullasch and Kaufmann)
claimed was that the peak intensities as related to
each other or to the Bremsstrahlung (the continuum)
did not correspond to the values one would expect from
the ion distributions known to be present in living
cells. They found, for instance, much too much Na
and Cl in their cells, and they did not believe that
this came from the effect Dr. Schimmel had just discussed.

SCHIMMEL thought that the result <u>could</u> come from the
different matrices, since the standards described by
Gullasch were used for the calculation of the content

of Na and K, and such standards are not applicable
for dried cells.

GULLASCH replied that they did not compare the standards with the air-dried cells; they only compared the
different peak heights.

LÄUCHLI (from the chair) suggested a postponement of
further discussion of this point.

[Editorial comment added later: We have given a full
paraphrasing of the interchange above because the
point is important, and it is our intention to render
the discussion in full. But the line of argument was
confusing, evidently because of a misunderstanding between the protagonists as to what each was claiming.
We therefore present a brief recapitulation here:
Dr. Schimmel correctly asserted that you cannot determine the relative amounts of Na and K in a dried
specimen simply by reference to data from a hydrated
standard, because absorption will affect the ratio of
the heights of the Na and K peaks, the Na radiation
being more strongly absorbed, and the absorption may
be very different in dried and hydrated material.
(Whether absorption will actually be much different in
any given case depends on beam voltage, specimen thicknesses, and takeoff angle, as well as on the local
specimen compositions). However Drs. Kaufmann and
Gullasch did not refer back to their hydrated standards in the interpretation of the spectra from the
dried cells; mainly they compared spectra taken from
regions outside of and within a single dried cell, a
procedure not subject to the criticism raised by Dr.
Schimmel.

At any rate the magnitude of the absorption effect
should not be underestimated. Discussion of this
point continued after the conference between Dr.
Schimmel and the editors, and various simple calculations showed that under some quite ordinary operating
conditions (not, of course, for ultrathin specimens),

differential absorption may easily change the observed ratio of Na and K peaks by a factor in the neighborhood of 1.3 — 1.5 when a given hydrated specimen undergoes dehydration].

F. INGRAM commented on the problem of the extracellular space. In the procedure which he described on the previous day, the extracellular space did not exist. This was the problem we always have if we are either freezing or drying, or plastic embedding. All the ions and electrolytes are deposited on the cell structures. It appeared to him that there might be a way to solve the problem by using cryoprotective agents such as glycerol or DMSO. Such agents might preserve the ion distributions in the extracellular space, as well as in the cells.

KAUFMANN disagreed, stating that it is known in electrophysiology that DMSO and glycerine drastically change the permeability of membranes even in the concentrations needed for cryoprotection. He doubted strongly that we can introduce such cryoprotective agents when we are trying to preserve the ionic distribution at the subcellular level, although such agents are very useful in the freeze etching technique and for the preservation of structural details.

M. INGRAM asked about the beetles which are able to withstand temperatures as low as approximately -40°C without freezing.

KAUFMANN replied that they have cryoprotective agents of a different sort, but nobody really knows what they are.

ECHLIN pointed out that even if you could use glycerol or DMSO as a cryoprotective agent, in a scanning microscope even at low temperatures it would still outgas, slightly at least, and such agents could give rise to severe contamination.

General Discussion

HALL went back to the earlier question, whether a front of freezing ice sweeps ions away. He referred to the work of Appleton, who had cut ultrathin frozen sections at low temperatures and had demonstrated, in frozen-dried sections of pancreas, that the intracellular electrolytes did not travel very far in the course of preparation (although the morphological appearance of the sections in the conventional electron microscope was rather poor). On the basis of the observed localisation, Appleton believed that the distributions of Na, K and Cl were well preserved within the cells. Admittedly, there was a lot of internal structure within these cells, and one must imagine that the electrolytes deposit on all kinds of membranes during the drying process. Hall went on to say that when frozen-dried sections were used as a means of preserving elemental distributions for the sake of microanalysis, the data from extracellular spaces were difficult to interpret partly because of the lack of microstructure there for the ions to settle on: the settling of ions on membranes must produce steep artificial gradients at the membranes, and prevent good measurements of the differences of concentrations between one side of a membrane and the other.

LÄUCHLI asked how Appleton's sections had been dried, and HALL confirmed that they were frozen-dried.

LECHENE noted that the problem of drying is very close to the problem of freezing, namely: What happens to the ions in the extracellular and intracellular compartments? He thought it likely that the intracellular compartments are in better condition than we realize, firstly because probably a large fraction of ions is bound, and secondly because it seems more and more probable that part of the water is organized, i.e. it is not in a liquid phase in which the ions are mobile. This amounts to something of a crystalline phase already in the living situation. But the real problem now, he judged, is to compare the intracellular concentrations (already measureable, according to his

210

argument) with the extracellular ones; he thought that
we must look for some matrix for the extracellular
space, which we can add as a suspension material,
which will prevent extracellular ions from attaching
to the outside of a cell during freezing or drying.

RUSS, returning to the concept of a high concentration
of ions being swept up to a membrane, asked whether
the diffusion rate of the ions varied with temperature.
He considered that the diffusion rate might be reduced
in the region of a cold membrane.

ECHLIN asked whether the increase in viscosity occa-
sioned by the decrease in temperature might not be
more effective in slowing down the movement of the
ions.

KAUFMANN considered that as long as there were liquid
and solid phases during freezing, there would be a
place in which ions could move. Even with freezing
rates of 1000-5000 °K/sec the ions could move, albeit
a small distance, in the cell.

SPURR judged that at -30°C, ions no longer exist as
such but are present as precipitated salts.

LÄUCHLI then raised the problem of the curling of sections of tissue cut on the ultracryomicrotome.
HUTCHINSON said that the best way to uncurl a section
was to use a hair on the end of a matchstick. He
judged that only the first five sections cut from a
specimen were any good — beyond that the ice damage
in the block was too severe. A warning was given that
frozen-hydrated sections can dry out within an hour
of storage in the cryochamber of the Sorvall instrument.

HUTCHINSON gave details of how he measured the beam
size and intensity using a Faraday cage. He cut down
background radiation by coating everything with epoxy
resin, and he suggested that carbon grids would reduce

background even further. RUSS suggested that the aluminum-coated nylon films used by Echlin and co-workers were satisfactory, as were the metal grids which had a single, central elliptical hole.

MACROCYCLIC POLYETHER COMPLEXES WITH ALKALI ELEMENTS IN EPOXY RESIN AS STANDARDS FOR X-RAY ANALYSIS OF BIOLOGICAL TISSUES

ARTHUR R. SPURR
Department of Vegetable Crops,
University of California, Davis, California

INTRODUCTION

Among the newer synthetic complexing agents, the macrocyclic polyethers have the unique property of making alkali metal salts soluble in nonpolar solvents (Pedersen, 1967). This capability was investigated as a basis for introducing ions such as Na^+ and K^+ into epoxy resin formulations with the intention of using the castings, or thin sections of them, as standards in X-ray analytical studies of biological tissues.

In conventional microprobe analysis the standards are commonly pure metals, formulated glasses, minerals, or chemical salts. In most of these, the matrix is often markedly different in mass, composition, and concentration of the element of interest from that of biological tissues. Consequently, in quantifying microprobe analyses, large correction factors are in principle required. Because of many variables, such corrections are seldom attempted and many analyses are presented in relative terms for biological systems.

Hall(1968) emphasized that one of the major problems in biological microprobe work is quantitation. He and his associates devised both relative and absolute methods primarily based on interpretations of mass as revealed by characteristic X-radiation and by

213

white X-radiation.

Numerous organic standards have been devised for
probe analysis, and it is appropriate to review some
of them briefly. Robison and Davis (1969) and
Andersen and Hasler (1965) used compressed pellets of
organometallic compounds to advantage, and the latter
emphasized the need for new standards, especially for
the lighter elements. Andersen and Hasler (1965) referred to another pelleting technique developed by
G. V. Alexander, in which freeze-dried aqueous solutions of organic compounds and salts were used as
standards in X-ray fluorescence studies. Hodson and
Marshall (1971) prepared Na and K standards formed on
carbon-coated cover slips to which controlled amounts
of collodion and alkali hydroxides in ethanol were
added. Birks (1969) referred to standards developed
by A. J. Tousimis in which metal compounds were suspended in plastic films. Ingram and Hogben (1968)
devised Na and K standards in frozen albumin which was
then treated like tissue specimens. They also referred briefly to their use of Cl-epoxy resin standards.
Membrane filters impregnated with K and then freeze-
dried were used as standards by Sawhney and Zelitch
(1969), and Rosenstiel *et al*. (1970) used sections
of methacrylate and agar-agar with Na, K, and Cl as
standards.

Although no artificial organic standards can fully reflect the elemental composition and distribution
within cells, a suitable organic matrix would nevertheless approximate tissue more closely than do the usual
standards used in microprobe analysis. Andersen (1967)
summarized the desired qualities of an organic standard indicating that it must be well defined chemically,
homogeneous at the micrometer level, and tolerant of
high vacuum and the heat imposed by the probe. Among
additional qualities that should be specified in view
of the high resolution of present-day instruments, the
homogeneity of the standard should be at the molecular,
rather than the micrometer level, with no aggregations

of the element of interest other than known stochio-
metric associations. Furthermore, if the tissue under
analysis is sectioned material, then the standard
should be amenable to sectioning at the same thickness.
In practice, this means that sections of the standard
10-50 nm to several micrometers or more may be needed
depending upon the instrument and the mode of analysis.

MATERIALS AND METHODS

Elemental Analysis

Two instruments were used in measuring the elements:
a) an Applied Research Laboratories electron
probe X-ray analyzer, model EMX-SM, operated at 15 keV
with a 5.0 µm beam diameter, with a sample current of
0.06 µA on brass, and an ADP analyzing crystal for K;
b) a Cambridge Scientific Instruments Stereoscan S4 operated at 10 keV at a magnification of X100,
and interfaced with a Kevex-ray solid state detector
and Qanta Metrix EDX/80 energy dispersive X-ray analysis system.

Transmission Electron Microscopy

Thin sections of the epoxy resin-KCNS-dicyclohexyl
[18]crown-6 complex standards were viewed on an RCA
EMU-3G operated at 50 keV. The sections, on 200 mesh
grids and not supported by an underlying film, appeared homogeneous and without any obvious substructure
at a magnification of about ×32,000.

Nomenclature and Structure of the Cyclic Polyethers

The cyclic polyethers used in this study (Figure
1) are referred to by abbreviated names as benzo[15]-
crown-5 and dicyclohexyl[18]crown-6 (Pedersen, 1967).
The first compound is more precisely identified as 2,
3-benzo-1,4,7,10,13-pentaoxacyclopentadeca-2-ene and

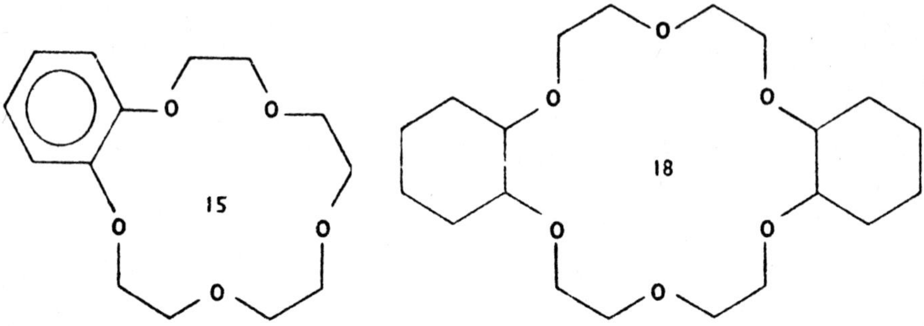

A. R. Spurr

Benzo 15 crown 5   Dicyclohexyl·18 crown 6

Figure 1. *Structural formulas of two cyclic polyethers.*

the second as 2,5,8,15,18,21-hexaoxatricyclohexacosane. The unusual ability of these neutral compounds to form alkali element-polyether complexes is related to the ion-dipole interaction between the cation and the negatively charged oxygens of the polyether ring. The site of the cation within the polyether ring in complexes of this type was first confirmed by Bright and Truter (1970) through X-ray crystallographic analysis.

Preparation of the KCNS-dicyclohexyl[18]crown-6-complex-epoxy Resin Standards

The KCNS-dicyclohexyl[18]crown-6 complex was introduced at room temperature directly into a freshly prepared low viscosity epoxy resin embedding medium (Spurr, 1969). A graduated series was prepared by adding appropriate amounts of the KCNS-cyclic polyether complex to the resin to achieve the following levels of K in milliequivalent/liter (meq/l): 200, 400, 600, 800, 1000, and 1200. The finely powdered complex was added to 1 g quantities of resin in small vials and, by stirring, it dissolved within 15 min. To insure homogeneity, the mixture was then warmed in a 70°C oven for not more than 4 min, and stirred again. After this it was promptly poured into embedding molds such as BEEM or gelatin capsules, and polymerized at 70°C for 16 h. This timing is not critical, and the

castings can be removed in 8 h or remain in the oven
for 24 h or longer.

*Preparation of the NaCNS-dicyclohexyl[18]crown-6
Complex*

NaCNS-cyclic polyether complexes are not available from commercial or research sources, but they
can be prepared by a straightforward method of
Pedersen and Frensdorff (1972). Among several possible procedures, the one followed in this study involved dissolving the ingredients in a common solvent,
methanol. Accordingly, 1:1 stoichiometric quantities
of NaCNS (0.81 g) and dicyclohexyl[18]crown-6 (3.725 g)
were added to 10 ml methanol. The mixture was transferred to an evaporating dish and much of the methanol
was evaporated by gentle heating. The residual methanol in the product was then removed in a vacuum oven
at 70°C for 16 h. The final product was a pale yellow,
semicrystalline, pliable mass.

*Preparation of the NaCNS-dicyclohexy1[18]crown-6
Complex-epoxy Resin Standards*

Procedures used in the preparation of the K standards were also used in the preparation of the NaCNS-
cyclic polyether complex-epoxy resin standards. In
both sets of calibration mixtures, the castings become
light amber to deep red during polymerization, but
this does not interfere with their use. The degree of
color is greater as the amount of complex in the resin is increased.

*Preparation of Complexes of Salts and Cyclic Polyethers
Directly in Epoxy Resin*

One of the methods used be Pedersen and Frensdorff
(1972) in the preparation of complexes of salts and
cyclic polyethers was to bring the reactants together
in a medium which dissolves one of them. The cyclic
polyethers used in this study are soluble in Spurr's

217

(1969) epoxy resin; thus it could serve as the reacting medium. Attempts to dissolve a wide range of common laboratory Na and K salts in the medium alone were unsuccessful. Although analytical methods were not used to evaluate the degreee of solubility of the salts, it was evident when mixtures of the resin and salt were observed under a microscope that most crystals retained their form. Nonionic neutral organic fluids cannot ordinarily be used for a range of calibration standards for microprobe analysis.

By introducing the cyclic polyether into the resin first, the capacity of the resin to dissolve Na and K salts, particularly at low cencentrations on the order of 100 and 200 meq/l was enhanced. Higher levels of salt were difficult to prepare, or could not be prepared in any practical way. The preparations were usually intractable sticky mixtures, particularly with salts such as NaCNS and KCNS. Possibly waters of hydration in such salts aggravated the problem. Complexes of NaI with benzol[15]crown-5 and of KI with dicyclohexyl[18]crown-6 could be made with greater ease, but still at only low levels of salt. From these experiences it was evident that the best method was to form the complex in a mutual solvent such as methanol, and then, after retrieving the complex, introduce it into the resin at the concentrations desired.

RESULTS AND DISCUSSION

Quantitative Analysis of Known Organic Standards for K Based on a Mineral Standard

The electron probe analysis of K (Kα) in a series of organic standards, the KCNS-dicyclohexyl[18]crown-6 complex in epoxy resin, is presented in Table I. Since the organic standards were formulated to certain specifications, the determination of their K content, by using a well-defined mineral standard, orthoclase,

TABLE I

ELECTRON PROBE ANALYSIS FOR K (Kα) OF ORGANIC STANDARDS COMPARING THE COMPOSITION BASED ON A MINERAL STANDARD VS. KNOWN FORMULATIONS

MINERAL STANDARD Orthoclase[a]		FORMULATED ORGANIC STANDARDS KCNS-dicyclohexyl[18]crown-6 complex-epoxy resin			
Composition K (Kα) ppm K	X-ray counts[b]	Composition meq/1 K	Composition ppm K	K (Kα) X-ray counts[b]	Composition based on Mineral standard ppm K
(Cm)[c]	(Im)[c]			(Is)[c]	(Cs)[c]
123,800 (12.38% wt.)	79,511	Control	0	1,229	--
		200	7,820	10,397	16,188
		400	15,640	16,189	25,206
		600	23,460	24,179	37,647
		800	31,280	27,873	43,398
		1000	39,100	35,411	55,135
		1200	46,920	39,060	60,817

[a]A well-analyzed standard, sample no. 5-168, obtained from Dr. B. Evans, University of Washington, Seattle, Washington.

[b]X-ray counts are the means of 10 determinations corrected for background and each determination was based on an integrated beam current for ca. 20 s.

[c]Symbols for mathematical relationship; see text.

provided information for judging the validity of using a standard that differs markedly in physical properties from the specimen. In calculating the K concentrations based on the mineral standard, use was
made of Castaing's first approximation, the *Intensity
sample/Intensity standard* ratio (Keil, 1967). The
particular mathematical relationship used was Cs =
Cm(Is/Im) where Cs is the potassium composition of the
organic standards, Cm is the potassium composition of
the mineral standard, Is is the K (Kα) X-ray intensity
of the organic standards, and Im is the K (Kα) X-ray
intensity of the mineral standard. The only corrections made were for the background. By inspecting the
ppm K values obtained in the last column of Table I,
it is evident that they are larger than the known formulated composition of the organic standards. A linear regression analysis calculated for a straight line
of the data in the last column resulted in a correlation of 0.98 whereas a perfect straight line or fit
equals 0.99. These results are almost as good as
those achieved in microprobe analysis of highly defined geological materials (Gulson and Loevering,
1968). The relatively high levels of X-ray counts
arising from the organic standards should be noted.
At the highest level of K, for example, the counts
were about half of those on the orthoclase; yet the K
content in the orthoclase was about 2.6 times greater.

Calibration Curve for K in Organic Standards

A calibration curve of the electron probe analysis for K (Kα) in prepared epoxy resin standards containing known amounts of the KCNS-dicyclohexyl[18]-
crown-6 complex is shown in Figure 2. The original
data plotted as points indicate the relationship of
X-ray counts to the known elemental concentrations.
The least squares regression analysis of the data, Y
(X-ray counts) on X (concentration of K), gave the
equation, Y = 6279 + 29.24X, which was significant at
the 10% level, with an R^2 value of 0.57. The trans-
formation of the equation to X = 0.0342Y - 214 enables

the prediction of the concentration of K(X) from the
X-ray counts (Y) of unknown samples. The original data
do not deviate appreciably from linearity in the range
of 200–1200 meq K/l, and this calibration segment
could probably be extrapolated to a reasonable degree
with confidence.

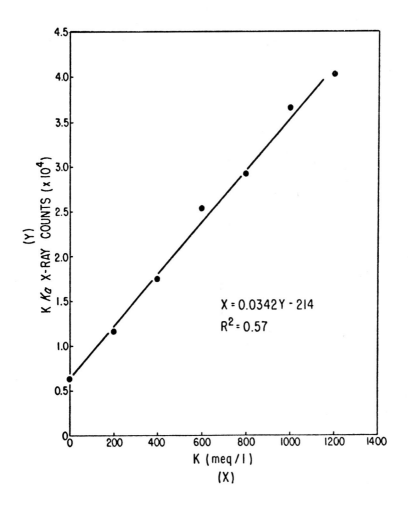

$$X = 0.0342\,Y - 214$$
$$R^2 = 0.57$$

Figure 2. *Linear, least squares fit calibration
curve for K in KCNS-macrocyclic polyether complexes
in epoxy resin standards.*

X-ray Energy Dispersive Analysis of NaCNS-cyclic Polyether Complex-epoxy Resin Standards.

A scanning electron microscope–energy dispersive system for the qualitative analysis of a NaCNS-dicyclohexyl[18]crown-6 complex in epoxy resin with 1200 meq/l Na was used in forming the spectrum shown in Figure 3. The spectrum, representing the X-ray energies for the first quarter of the periodic table, has

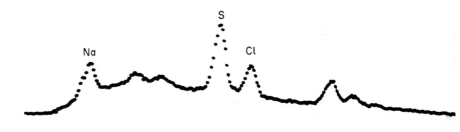

Figure 3. *Oscilloscope record of a 200 s. X-ray energy analysis spectrum of an epoxy resin standard containing NaCNS-dicyclohexyl[18]crown-6 with 27,588 p.p.m. Na(1200 meq Na/1). Note peaks for* Na *(1.04 keV),* S *(2.31 keV), and* Cl *(2.62 keV). The energy range for this spectrum is 0.7 keV to 3.28 keV with 10 eV channel separation.*

a prominent peak for Na at 1.04 keV, and also well-defined peaks for S and Cl at 2.31 keV and 2.62 keV, respectively. There are also two minor peaks between Na and S representing Al and Si. The peaks at the right were not identified. Both the Na peak and the prominent S peak relate to the NaCNS introduced into the epoxy resin through the cyclic polyether complex, as these peaks were not present in the spectrum from the control sample. The Cl peak is a consequence of the epichlorohydrin used in the manufacture of the D.E.R. 736 (diglycidyl ether of polypropylene glycol),

the flexibilizer in the epoxy resin (Spurr, 1969).
The hydrolyzable Cl content of D.E.R. 736 averages
0.5% and, after its incorporation with the other components of the resinous mixture, the estimated level
of Cl is about 0.07%.

X-ray counts for Na, S, and Cl were also made for
a series of epoxy resin standards of known Na composition, but a good linear fit was not achieved in the
initial tests. A K series based on a KCNS-cyclic polyether complex in epoxy resin was also attempted, but
the conductive silver paint used in securing the samples to the stubs introduced prominent L-line emission
energies in the region of K (Kα) at 3.31 keV.

In this initial study the epoxy resin standards
used for analysis were thicker than the depth of penetration and zone of excitation by the electron beam
in both the SEM-energy dispersive system and the conventional electron probe system. Sections of the
standards were easily cut with glass knives on a
Cambridge ultramicrotome at thicknesses ranging from
about 50 nm to 8 μm, and this range could probably be
extended in either direction. As the thin sections
had no obvious substructure when viewed under TEM,
this alkali element-cyclic polyether complex-epoxy
resin system of calibrated organic standards seems to
meet many of the requirements for organic standards
in the probe analysis of biological tissues. Several
aspects of this system need further development and
evaluation. Information needs to be obtained on the
stability of the system, and especially of the complexed cation, under the variety of conditions and modes
encountered during analysis. It should be determined
if elements in the sections of the standards will be
leached by the water in the trough before the sections
are retrieved. Since most alkali element-cyclic polyether complexes are soluble in nonpolar solvents, this
is not likely to be a problem; but if it should arise,
it would be possible to use other fluids in the trough
or to cut sections dry by cryomicrotomy.

With further experience in the use of the standards, correction procedures should be introduced to accommodate for those properties of the standards that do not closely approximate biological tissues. Among the complications that may arise, particular consideration should be given to the absorption of ultra-soft radiation. Variations in tissue density and the diversity and localization of elements in biological samples may also affect such phenomena as fluorescence and the depth of beam penetration.

The approach taken in this study may be extended to other elements. In addition to Na^+ and K^+, the following cations listed by Pedersen and Frensdorff (1972) might be introduced into epoxy resins for use as organic standards: Li^+, Rb^+, Cs^+, NH_4^+, RNH_3^+, Ag^+, Au^+, Mg^{2+}, Ca^{2+}, Sr^{2+}, Ba^{2+}, Ra^{2+}, Zn^{2+}, Cd^{2+}, Hg^+, Hg^{2+}, La^{3+}, Tl^+, Ce^{3+}, and Pb^{2+}.

Since each salt used in forming a complex conveys an anion into the system, it should be possible to prepare standards for an even wider variety of elements. Thus, the thiocyanate and iodide anions introduced into the standards via their Na and K salts should permit these standards to serve in quantitative determinations for S and I, respectively. The Na or K salts of Br^-, MnO_4^-, $CoCl_4^{2-}$, and $Fe(CN)_6^{4-}$, are other possibilities. However, cyclic polyether complexes with salts of high lattice energy such as the phosphates and fluorides cannot be prepared (Pedersen, 1967).

The principle of using neutral complexing agents in epoxy resins can perhaps be extended to other compounds to further widen the range of standards. The potential use for the macrocyclic polyether complex systems developed in this study seems to be considerable, and it is recommended that they be tested widely in X-ray analytical research.

Macrocyclic Polyether Complexes

CONCLUSIONS

The unique property of the macrocyclic polyethers to form complexes with Na and K permitted the formulation of organic standards for use in X-ray analytical studies of biological tissues.

An adventageous method for the preparation of complexes of NaI, KI, NaCNS, and KCNS with the cyclic polyethers is to form the complexes in a mutual solvent such as methanol. The solvent is then removed by gentle heating in a vacuum oven.

The complexes can be introduced into Spurr's low-viscosity epoxy resin in concentrations up to 1200 meq/l or perhaps more. Polymerization of the resin-complex mixtures results in homogeneous castings. Depending upon the instrumentation and mode of X-ray analysis the castings themselves can be used as standards or they can be sectioned at any thickness.

ACKNOWLEDGMENTS

It is my pleasure to express appreciation to Professor M. R. Truter, Agricultural Research Council, University College, London, UK, and Dr. H. K. Frensdorff, E. I. du Pont de Nemours and Company, Wilmington, Delaware, for providing macrocyclic polyethers for this study. I am also grateful to Dr. K. N. Paulson and Mr. B. Lathrop of the Department of Vegetable Crops and Mr. R. Wittkopp of the Department of Geology, University of California, Davis, for their assistance.

REFERENCES

Andersen, C. A., 1967, *Methods of Biochem. Anal.*, 15, 147-270.

225

Birks, L. S., 1969, *X-ray Spectrochemical Analysis*, Interscience Publishers, New York.

Bright, D., and Truter, M. R., 1970, *J. Chem. Soc.*, B 1970, 1544-1550.

Gulson, B. L., and Loevering, J. F., 1968, *Geochim. et Cosmochim. Acta.*, 32, 119-122.

Hall, T., 1968, In: *Quantitative Electon Probe Microanalysis*, Proc. Seminar, National Bureau Stds. Gaithersburg, Maryland, Heinrich K. F. J., Ed., 269-299.

Hodson, S., and Marshall, J., 1971, *J. Microsc.*, 93, 49-53.

Ingram, M. J., and Hogben, A. M., 1968, In: *Developments in Applied Spectroscopy*, 6, 43-54.

Keil, K., 1967, *Fortschr. Miner.*, 44, 4-66.

Pedersen, C. J., 1967, *J. Am. Chem. Soc.*, 89, 7017-7036.

Pedersen, C. J., and Frensdorff, H. K., 1972, *Angew. Chem. Internat. Ed. Engl.*, 11, 16-25.

Robison, W. L., and Davis, D., 1969, *J. Cell Biol.*, 43, 115-122.

Rosenstiel, A. V. P., Höhling, H. J., **Schnermann, J.**, and Kriz, W., 1970, *Proc. 5th Natl. Conf.*, *Electron Probe Analysis*, Electron Probe Analysis Society of America, New York City. 33A-33B.

Sawhney, B. L., and Zelitch, I., 1969, *Plant Physiol.*, 44, 1350-1354.

Spurr, A. R., 1969, *J. Ultrastr. Res.*, 26, 31-43.

Macrocyclic Polyether Complexes

DISCUSSION

QUESTION: Can the actual composition of the organic standards be extended?

ANSWER: We can easily lower the levels of cations in the standards and can probably extend the range higher. There are a number of alternative formulations that haven't been examined as yet.

QUESTION: What happens to the anion?

ANSWER: In the studies of Pedersen the anion is reported to hang right in there. It is not regarded as part of the complex yet it remains closely associated in stoichiometric relationships such as 1:1 or 2:1, depending upon the salts.

QUESTION: Are the cyclic polyethers commercially available?

ANSWER: The dicyclohexyl[18]crown-6 and the benzo[15]-crown-5 are commercially available. A few complexed forms may be available as research chemicals.

QUESTION: Can one use just one complex to set up organic standards?

ANSWER: I think that the dicyclohexyl[18]crown-6 is the one to center one's attention on if you are interested in sodium and potassium standards. Other cyclic polyethers could accommodate larger cations.

QUESTION: Do you see any possibilities for chlorides?

ANSWER: Cyclic polyether complexes with alkali salts having small anions such as chloride are not as easily prepared as those with large anions, e.g. bromide, iodide, and thiocyanate. I should emphasize that the complexes are neutral compounds which can be put into an epoxy resin system so that the cation of interest is dispersed uniformly.

QUANTITATIVE ANALYSIS OF THIN SECTIONS, AND THE CHOICE OF STANDARDS

T. A. HALL AND P. D. PETERS
*Cavendish and Biological Microprobe
Laboratories, Cambridge, England*

In this paper we compare the use of organic and mineral standards for the quantitative microprobe analysis of thin tissue sections, especially in conjunction with the continuum method of quantitation (Hall *et al.*, 1972; Hall and Werba, 1971; Hall, 1971; Hall *et al.*, 1973). When this method is used, the chief requirements imposed on the standards are that they must contain the elements to be assayed, and that they must be thin, homogeneous, and of known composition and known stability under the electron beam.

Suitable organic standards are usually prepared by adding salts of the relevant elements to organic material, mixing thoroughly and sectioning. For example, standards for Na, Cl or K have been prepared by adding electrolyte salts to tissue homogenates (Kriz *et al.*, 1972) or to albumin-agar (Gehring *et al.*, in press) or to 20% albumin (Ingram and Hogben, 1968). One advantage of such organic standards is that they are quite similar in composition to specimens of soft tissue, so that quantitation can be carried out almost empirically, usually by working on the linear region of a calibration curve, without reliance on any elaborate theory. A disadvantage is that it may be difficult to achieve homogeneity. A further feature is that the composition of organic standards is altered by the electron beam — whether this should be regarded as an advantage or a disadvantage will be considered briefly below.

Probably the simplest way to prepare <u>mineral</u> standards is by grinding and then pipetting from a suspension of ground material onto an ultrathin plastic supporting film. However, it is difficult to control the particle thickness in a ground preparation, so that incalculable absorption effects may occur, especially if soft X-rays must be observed or the mineral contains atoms of high atomic number (for example, FeS_2 as a standard for sulphur). Therefore it is better to embed the ground particles and to cut ultrathin sections of them. Mineral standards are homogeneous and the composition is usually well known. Many minerals are reliably stable under the beam (cf. Sweatman and Long, 1969) so long as one takes care not to irradiate with too great a current per unit area — i.e., one should avoid focussing to the ultrafine spots now available in SEM's or EMMA instruments. A current of about 5 nA/μm^2 usually provides ample signal.

In the comparison of mineral and organic standards two major points arise, related to the theory of quantitation and to beam damage. In contrast to the position with organic standards, when mineral standards are used for quantitative analysis with the continuum method, one must resort to an elaborate theory of quantitation, whose range of validity is still under investigation. The main reason for this is that the mean atomic number is almost the same in soft tissues and in organic standards, but is usually much higher in mineral standards, and in order to use the intensity of the continuum radiation as a measure of irradiated mass one must take account of the dependence of the efficiency of generation of continuum radiation on atomic number; the correct expression for this dependence is not yet precisely established.

With respect to beam damage, one may hope with an

organic standard that the degree of damage is the same in standard and specimen, so that the effect cancels and one obtains a good measure of the elemental mass fractions as they were prior to irradiation. But the degree of beam damage seems to depend on many factors, and it seems unwise to assume that specimen and standard are really affected in general to the same extent. In the present situation, until we learn how to avoid beam damage, it seems a matter of taste whether one prefers mineral standards, which should give valid measurements of the specimen as it actually existed during analysis, or organic standards which may give measurements closer to the concentrations existing prior to analysis.

In Cambridge we are trying to use the two types of standard together, hoping that the comparison will add to our understanding of both beam damage and the theory of quantitation. Here we shall describe two such comparisons of mineral and organic standards, which were intended respectively for measurements of the electrolyte elements K, Cl and Na, and for the measurement of sulphur.

The electrolyte measurements, which have been described elsewhere (Höhling *et al.*, in press), were carried out in unembedded dried frozen sections of rat kidney. Sets of organic standards were prepared by the addition of various amounts of salts to kidney homogenates. The mineral standards were sections of NaCl, sanidine (85 moles $K_2O \cdot Al_2O_3 \cdot 6SiO_2/15$ moles $Na_2O \cdot Al_2O_3 \cdot 6SiO_2$) and albite (99 moles $Na_2O \cdot Al_2O_3 \cdot 6SiO_2/1$ mole $K_2O \cdot Al_2O_3 \cdot 6SiO_2$). The test of standards takes the form of a comparison of two independent ways of measuring the electrolyte concentrations in the homogenates, by flame photometry and by microprobe analysis, the mineral sections serving as the standards in the latter case. Since the organic standards were internally consistent in that they gave a calibration curve which was a straight line through the origin, and the three mineral standards were internally

consistent with respect to their common element Na
(cf. Hall and Werba, 1971), the results can be summarized in the simple form of Table I.

TABLE I.

A Comparison of Organic and Mineral
Electrolyte Standards

Organic standard: kidney homogenate
Mineral standards: sanidine, albite, NaCl

Element	Mass fractions,% I	II	Ratio, I/II
Na	0.55	0.93	0.59
Cl	0.53	0.98	0.54
K	0.86	1.47	0.59

Average 0.57

Column I: Dry weight mass fractions in
unprobed homogenate, by flame photometry.
Column II: Dry weight mass fractions in
probed sections of homogenate, deduced from
sections of mineral. The average ratio 0.57
implies a mass loss of 43%, of which approximately 15% is attributable to a slow thermal effect. (Data of Delft group, 2/1971).

The flame photometric measurements were carried
out of course on aliquots of homogenates which were
not exposed to the electron beam. Thus, Table I indicates that the electrolyte concentrations in the
homogenate sections during microprobe analysis were
higher than the concentrations in the unprobed aliquots by a factor of approximately 100/57. Our interpretation is that the homogenate sections lost approximately 43% of their mass under the electron beam
(perhaps 30% from the rapid rupture of chemical bonds
and the remainder from a slower rise in temperature).
The mineral standards and the associated theory of
quantitation seem to give a valid assay of the specimen as it exists under the beam, and the suggested

Quantitative Analysis of Sections

loss of mass in the organic standards is consistent with other data on similar material (cf. the paper "Measurement of mass loss ..." by Hall and Gupta in this volume).

In order to discuss our recent observations with the _sulphur_ standards, we must first recapitulate the basic equations used for quantitation. X-ray theory leads to the approximate equation

$$R_1/R_2 = G_1/G_2 \qquad (1)$$

where R is the ratio of the recorded counts of a characteristic line of a particular element "x" to the recorded counts of the continuum radiation, $G = N_x/(\sum_r N_r Z_r^2)$ where N_x is the number of atoms of the element "x" per unit volume and the sum is extended over all constituent elements, Z is atomic number, and subscripts 1 and 2 identify probed microvolumes (in specimens and/or standards). In the present case, where we are comparing two standards of known composition, G_1 and G_2 can be calculated from the given compositions, so equation (1) gives a prediction which can be checked against the value of R_1/R_2 which comes directly out of the experimental data.

For sulphur analysis we use the mineral standard FeS_2 (sections provided by H. J. Höhling) and our organic standards consist of Araldite with a sulphur compound added (sections provided by H. Jessen). The data are summarized in Table II.

According to the upper part of the Table the measured ratio R_1/R_2 is 5/4 of the predicted value, which is understandable in terms of a loss of 20% of the mass of the Araldite under the electron beam. The lower part of the Table summarizes a later run with the Araldite standards alone, setting up a calibration curve. According to the later data the standards are consistent under the assumption of no loss of mass of the Araldite. The question will have to be resolved

TABLE II

A Comparison of Organic and Mineral
Standards for Sulphur

Organic standard (1): Sulphur in Araldite (H. Jessen)
Mineral standard (2): Sectioned FeS_2 (H. Höhling)

Test: $R_1/R_2 \overset{?}{=} G_1/G_2$ ($R = $ char./white ; $G = N_x / \sum NZ^2$)

$G_1 = 0.000344$ $G_2 = 0.00215$ $G_1/G_2 = 0.160$

Run	R_1	R_2	
1	0.0190	0.092	
2	0.0204	0.104	
3		0.098	
Average	0.0197	0.098	$R_1/R_2 = 0.201$

(estimated residual mass = 160/201 = 80%)

Later run 0.0155 R_1/R_2 (0.0155/0.098) = 0.158
(estimated residual mass = 160/158 = 101%)

by repetition of these very recent data. At the moment it seems at least that equation (1) is not too inaccurate, even for a mineral standard where most of the continuum radiation comes from an element (iron) with atomic number as high as $Z = 26$.

In conclusion, mineral and organic standards are both useful in the quantitative analysis of sections of soft tissue, and it is helpful to use the two types of standards together. At present the discrepancies between them are understandable on the basis of beam damage and the limited precision of the theory of quantitation.

REFERENCES

Gehring, K., Doerge, A., Nagel, W., and Thurau, K., *Proc. Internat. Symposium on Modern Technology in Physiological Sciences, Munich, July 1971,* in press.

234

Quantitative Analysis of Sections

Hall, T. A., 1971, In *Physical Techniques in Biological Research,* 2nd Edition, Vol. IA, G. Oster, Editor, Academic Press, New York and London, pp. 157-275.

Hall, T. A., and Werba, P., 1971, In *Proc. 25th Anniversary Meeting of EMAG,* Institute of Physics, London and Bristol, pp. 146-149.

Hall, T. A., Röckert, H. O. E., and Saunders, R. L. deC. H., 1972, *X-Ray Microscopy in Clinical and Experimental Medicine.* Charles C. Thomas, Springfield, Illinois, U.S.A.

Hall, T. A., Anderson, H. C., and Appleton, T. C., 1973, *J. Microscopy,* 99, 177-182.

Höhling, H. J., Hall, T. A., Kriz, W., von Rosenstiel, A. P., Schnermann, J., and Zessack, U., *Proc. Internat. Symposium on Modern Technology in Physiological Sciences, Munich, July 1971,* in press.

Ingram, M. J., and Hogben, C. A. M., 1968, *Develop. Appl. Spectrosc.,* 6, 43-64.

Kriz, W., Schnermann, J., Höhling, H. J., von Rosenstiel, A. P., and Hall, T. A., 1972, In *Recent Advances in Renal Physiology,* Karger, Basel, pp. 162-171.

Sweatman, T. R., and Long, J. V. P., 1969, *J. Petrology,* 10, 332-379.

DISCUSSION

COMMENT: In reference to the two differing runs in Table II, was the same batch of Araldite used, and was there a long time (months perhaps) between the runs? Araldite may change slowly with polymerization over a period of months.

RESPONSE: In fact the first run was in the morning and the second was during the afternoon of the same day. But it seems quite possible that the Araldite may change in the vacuum of the column, or during irradiation in areas which are close to the probe though not directly under it.

COMMENT: It would be interesting to compare standards as you have done, but at low temperature.

RESPONSE: Yes. Unfortunately this program of observations is being carried out with our EMMA instrument, which is not equipped with a cold stage.

COMMENT: You indicated some doubt about Kramers' law, the proportionality of continuum intensity per atom to Z^2. Is this dependence worth more study? Is the law a summary of experimental observations or is it based on theory?

RESPONSE: It would definitely be worthwhile to study the Z^2 dependence more closely. Kramers deduced this dependence from a very crude theory, pre-dating quantum mechanics. In fact a Z^2 dependence can also be derived from a quantum mechanical theory if one assumes an unscreened Coulomb electrostatic field around each atomic nucleus, but of course this is only an approximation and the true dependence of continuum intensity on atomic number must be more complex. However, in practice the Z^2 law has given us good results, and it is pleasing that now it seems useful even for atomic numbers as high as $Z = 26$ (iron), as in the case of the FeS_2 standard, where most of the continuum radiation comes from the iron.

COMMENT: At this point we could use many more measurements of electron cross sections, carried out with thin films and Si(Li) detectors. Then we could carry out measurements in thin specimens entirely on the

Quantitative Analysis of Sections

basis of improved theory, with no need for standards.

RESPONSE: Yes, it is possible to work from theory in this way, so long as you have some kind of measure of specimen thickness, or if you want to measure only elemental ratios. I myself prefer rather to work with standards at the time when the specimen is analysed.

SECOND RESPONSE (from the chair): Of course you need standards in any case. There are really two ways of using standards: You can use them when the measurements are made on the specimens, or you can use them at another time to measure the parameters (like cross sections) which are needed for the application of a theory. Our experience is that the more standards you use, the more you are bound to come close to "the truth".

COMMENT: You seem to have a touching faith that the mineral standards are homogeneous, but our experience with microprobe analyses of minerals shows that they are often far from homogeneous.

RESPONSE: You certainly have to be careful in the choice of minerals and in their source. We have generally tried to use minerals which are recommended or shown to be favorable in the extensive review by Sweatman and Long (1969), and we fortunately are able to get most of our supply directly from Jim Long's laboratory. Also, when you are looking at transmission electron images of thin sections of small pieces, you can get a fairly good idea whether a particular piece is likely to be good or "dirty". In a run including many spots, the spread in characteristic/continuum ratios is higher than accounted for by counting statistics, and I suppose that inhomogeneity probably accounts for a large part of this surplus spread, but the probably error in the mean value of the ratio is still generally less than 10%, and is usually about 5%, I would guess.

EXPERIENCE OF DIFFRACTIVE AND NON-DIFFRACTIVE QUANTITATIVE ANALYSIS IN THE CAMECA MICROPROBE

W. A. P. NICHOLSON
Institute of Medical Physics,
University of Münster, Germany

INTRODUCTION

Our quantitative measurements have been based on the equations developed for thin sections by Hall (1968, 1971). The equation which is satisfactory for our analysis of soft tissue containing elements in concentrations of less than one or two percent, may be expressed:

$$C_x = R_x \, A_x \, G_x \, H$$

where C_x is the concentration of the assayed element x, of atomic number A_x, G_x is a factor calculated from the chemical formula of the standard used for x, and H is a function of the specimen matrix, which has a value of 3.28 for soft tissue. The factor R_x can be written:

$$R_x = \frac{(Characteristic \; count/Continuum \; count)_{specimen}}{(Characteristic \; count/Continuum \; count)_{standard}}$$

and is the quantity actually measured during the analysis (Figure 1). From the measured continuum or "white" counts must be subtracted the white counts originating from the specimen support film and other parts of the instrument. The background must also be

CONCENTRATION $C_x = A_x \, G_x \, H \, R_x$

$$R_x = \frac{(characteristic \ count/continuum \ count)specimen}{(characteristic \ count/continuum \ count) \ standard}$$

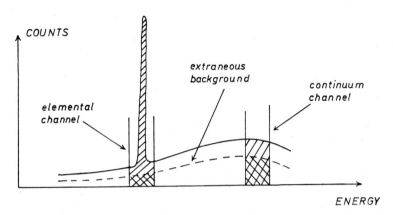

Figure 1. *The basis of the method of the quantitative measurement.*

subtracted from the characteristic peaks. The paper deals with the practical difficulties encountered in making these background subtractions, and gives some comparison between the roles of diffractive and non-diffractive analysis in a standard microprobe.

MEASUREMENT OF THE WHITE RADIATION

The majority of electron microprobes have been designed for the analysis of compact specimens, in which only the characteristic lines need to be measured. The only precaution usually taken against scattered electrons and x-rays is the use of deflecting magnets in front of the thin windows of long-wavelength spectrometers, to prevent stray electrons generating x-rays in the window. The scattered electrons generate continuum x-rays in the region of the specimen, and the intensity can be very high, particularly

240

in the case of microprobes in which it is possible to form electron microscope images. Figure 2 shows the region of the Cameca near to the specimen, when the electron microscope attachment is in place. The specimen causes scattering from the incident beam in both the forward and backward directions, which can strike

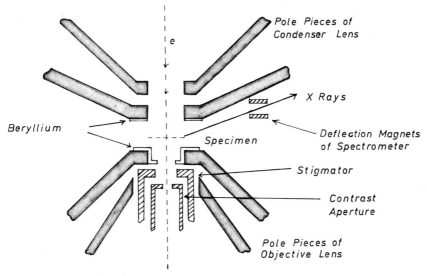

Figure 2. *The region around the specimen in the Cameca microprobe*

the nearby metal parts, such as the pole pieces, the stigmator plates and the contrast aperture. Some of the electrons generate x-rays, and others are re-scattered towards the specimen and its support where further x-rays can be produced. Measurement shows that, for a support film on a normal electron microscope grid in the Cameca probe before modification, about 95% of the white radiation is extraneous and only 5% is generated in the specimen by the primary beam. Figure 3 shows the measured intensities of the white and Cu Kα radiations on a thin formvar support film as a function of the distance from the grid bar. Even in the center of the mesh, the stray Cu radiation would prevent analysis of small concentrations of this element.

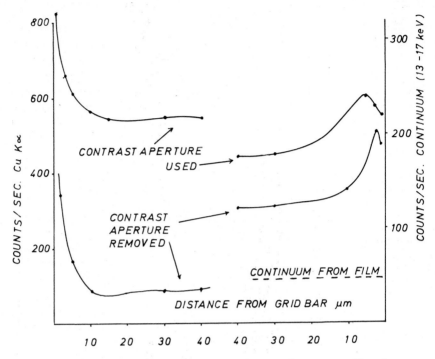

Figure 3. *The copper and continuum radiation levels measured on a Formvar film supported on a copper grid.*

It follows that the electrons generating the characteristic line will also generate a continuous spectrum in the metal. Estimates (Cosslett and Nixon, 1960) show that for 30 keV electrons striking copper, the characteristic line makes up about 30% of the total generated spectrum.

A further difficulty is that it is not possible to subtract a high background accurately, because the intensity of the stray radiation is approximately proportional to the mass in the beam. This arises because a higher mass produces more scattering of the primary beam, thus the background measured on a support film close to the specimen would give a (false) low value. The value for the film alone shown in the figure, was estimated from similar films supported on holders without grid bars.

To reduce stray radiation the standard 3 mm copper electron microscope grids have been replaced by 3 mm aluminum discs with a single 1.5 mm hole in the center, over which the specimen is supported on the usual film. The pole pieces have been shielded with beryllium and a beryllium aperture about 1 mm in diameter has been placed over the stigmators, as a scatter trap. Measurement of the stray radiation using a thin film on aluminum holders, showed that it is preferable to measure the specimen at points at least 100 μm from the metal support, so that grids can not be used. An important advantage is that the modifications permit measurements to be made with the contrast aperture in place. The background in the white channel from the specimen support film and stray radiation is typically 1/2 to 1/3 of the total measured from the specimen, and this fraction could possibly be reduced by thinner films and conducting coatings. With these signal-to-background ratios, quantitative measurements are possible.

When thicker (6 μm) sections are to be measured, and specimen images are formed by scanning, different specimen holders are used. The specimens are supported on a nylon film about 250 nm thick over an aluminum tube 10 mm in diameter. These specimen holders are then mounted with at least 100 mm of clear space beneath the open end of the tube, in the standard microprobe attachment. Measurements of the white radiation from these specimens and support films covered with an aluminum coating of 20 nm on both sides, showed that signal-to-background ratios of between 4:1 and 10:1 can be routinely achieved.

COMPUTING

The program used for measurements made with the crystal spectrometers simply substitutes the data which have been recorded on punched paper tape, into

the equations to be evaluated. An average value of a
number of measurements of the white radiation from the
support film, is subtracted from each measured value
of the white radiation from the specimen. For conven-
ience, a single value of the off-peak of the spectrometers may be recorded for a set of measurements. A
background value for each peak (characteristic line)
measurement is then calculated from this in proportion
to the white radiation of the two measurements. Since
the peak-to-background ratio in the characteristic
channel rarely falls below 30:1, this procedure is
quite accurate. For measurements made with the spectrometers then, the computor performs simple calculations which for a typical 2 hr measurement series
could be done by hand in about an hour.

The situation for spectra recorded with the silicon, non-diffractive detector is quite different. In
normal 6 μm sections, containing elements in concentrations of about one percent, typical peak-to-background ratios lie in the range of 3:1 to 1/3:1, and
thus the background subtraction must be made very
accurately. Figure 4 shows the basis of a simple
method that has been considered. The background is

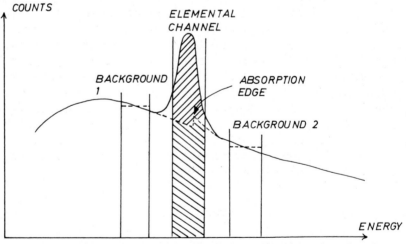

Figure 4. *A simple method of estimating background levels.*

measured on either side of the peak. The multi-channel analyser records the integrated counts in each channel and the average background is subtracted from the peak. It should be noted that this is the method that would have to be used if the detector were used with single channel analysers.

Except for measurements on standards which have high peak-to-background ratios, this method is not likely to give good results, because the continuum can not be assumed to be constant over the large energy range of the three channels. Other possible methods of estimating the background by hand calculation require the spectrum to be plotted out, and an extrapolation to be made from either side of the peak. Russ (1971) points out that background subtractions based on this method will give peaks that are too high, because the absorption edge of an element, which is not resolved from the elemental line, can alter the level of the background under the peak on the low energy side. A further difficulty arises in most cases because there are many peaks present in the spectrum, and the background must be extrapolated over a large energy range. Another method suggested by Russ is to strip the peaks from the original spectrum, leaving only the background. This is then subtracted from the original spectrum leaving the peaks. It is obvious that either of these methods require too much time to be performed by hand calculation. To record the data channel by channel from the multi-channel analyzer, would alone take several minutes per measured spectrum.

It would seem, therefore, that the principal role of the computor associated with non-diffractive detectors, is to perform accurate background substractions. Once this has been achieved the peak values can be substituted into the quantitative equations to calculate the concentrations without difficulty.

To examine these ideas in practice, a few measurements were made on specimens of known average elemental concentrations, these having been previously determined using crystal spectrometers. The spectra from apatite standards, predentine, and rat tail tendon specimens were recorded and stored in the EDAX 707 multi-channel analyzer. The analyzer gave the integral values of the peaks and background windows and the data were evaluated by hand using the simple method previously described. The spectra were then transferred to the EDIT mini-computer, which performed the background subtractions and integrated the remaining peaks. It is very difficult to evaluate the merit of the background subtraction routine used, but for the present it can be assumed to be much better than that of the hand calculation procedure. A comparison of the two sets of results, showed that the hand calculated values were too high by 5-10% for the standards, and between 10% too low and 30% too high for the specimens, when compared to the values obtained from the computer.

It is interesting to note that some small peaks were measured by the computer that could not be evaluated by hand calculation. More important, no calcium peaks were visible in the spectra from tendon and so could not be measured by either method. Calcium in rat tail tendon has been measured routinely, and found to be present in an average concentration of 0.05% (Nicholson, 1973). This concentration is clearly below the limit of detectability of the non-diffractive detector.

CONCLUSION

To draw some general conclusions from this, it is useful to compare typical measuring conditions in diffractive and non-diffractive spectrometry. For the crystal spectrometers, there is a practical limit on the line intensity, set by the maximum beam current that may be used without causing specimen damage, typically in the range 10 — 100 nA. Using the silicon

Analysis of the CAMECA Microprobe

detector, the limit is set by the maximum total count-rate for the whole spectrum that the detector can handle, about 10,000 c/sec. In the interface on the Cameca, the specimen-to-detector distance is 65 mm subtending a solid angle of 10^{-3} at the specimen. Even at this low collecting efficiency, about 5 nA beam current is enough to badly overload the detector, even using an ultra-thin specimen.

It is thus not possible to take advantage of the high efficiency of the silicon detector for the elemental line in a normal microprobe. The result is that the crystal spectrometer has an overall higher sensitivity.

These observations are in agreement with the calculations of Hall (1971), who estimated that the non-diffractive detector would have a higher sensitivity, only when the electron beam size were less than about 0.1 μm; and that for the normal electron microprobe with a beam diameter of 1 μm, the crystal spectrometer would be more sensitive.

REFERENCES

Cosslett, V. E., and Nixon, W. C., 1960, In *X-ray Microscopy*. University Press, Cambridge 231-232.

Hall, T. A., 1968, In *Quantitative Electron Probe Microanalysis.* (Heinrich K. F. J., Ed). National Bureau of Standards Special Publication No. 298, 269-299.

Hall, T. A., 1971, In *Physical Techniques in Biological Research.*, Academic Press, London, 1A, 2nd Ed., 203-206.

Nicholson, W. A. P., 1973, in press.

Russ J. C., and Barnhart, M. W., 1971, In *Proceedings*

W. A. P. Nicholson

of the Sixth International Conference on X-ray Optics and Microanalysis. (Shinoda, G., Kohra, K. and Ichinokawa, T., Eds.), University of Tokyo Press, Tokyo, 271-278.

DISCUSSION

COLEMAN: Did you find any potassium in rat tail tendon?

NICHOLSON: Yes, but as I had no potassium standard I can give no estimate of the concentration.

RUSS: Could it be a problem of potassium Kβ obscuring the calcium Kα?

NICHOLSON: Even after 300 sec measuring time the potassium Kα peak was small. The potassium Kβ and the calcium Kα peaks were both lost in the background.

HEINRICH: Are the absorption jumps at the edges really so important in biological matrices?

NICHOLSON: I don't think they usually have much effect.

RUSS: I think the absorption edges could be important if you were measuring sodium and magnesium in bulk tissue, when the penetration depth was large. In measurements of elements like calcium in thin sections the effect would be unimportant.

QUANTITATIVE ANALYSIS OF BIOLOGICAL MATERIAL USING COMPUTER CORRECTION OF X-RAY INTENSITIES

R. R. WARNER[1] AND J. R. COLEMAN
Department of Radiation Biology and Biophysics
University of Rochester School of Medicine
and Dentistry, Rochester, New York

INTRODUCTION

There are essentially three techniques presently available for microprobe quantification of biological material. One technique involves the use of specially prepared standards. Another technique utilizes a theoretical development based on the local variation of mass per unit area commonly observed in biological specimens. The third technique, and the one which will be discussed at length in this paper, is based on the conventional theory of quantitative analysis as developed in the metallurgical and mineralogical fields.

All of these quantitative techniques utilize the theoretical backgound for quantitative microanalysis as established by Castaing (1951) in his dissertation. Following this technique, the mass concentration of element A in a complex specimen, C_A, is to a first approximation equal to the measured x-ray intensity from A in the specimen, I_A, divided by the intensity from a pure elemental standard, $I(A)$:

$$C_A = I_A/I(A) = K_A \qquad (1)$$

---

[1]Present Address: Dept. of Physiology, Yale University School of Medicine, New Haven, Conn.

The use of this specimen-to-standard x-ray intensity
ratio provides internal compensation for the various
and variable instrument parameters affecting the x-
ray counting rate.

Since the measured intensity of element A is a
complex function of both the physical properties and
the chemical composition of the specimen, and since
the chemical composition and physical properties of
specimen and standard are rarely identical, the relationship given by equation (1) may lead to a poor or
even misleading approximation. A common technique in
biological microprobe analyses which avoids this problem (as well as avoiding the complexities of theoretical corrections to be discussed later) involves the
preparation of artificial standards with composition
and physical properties similar to the unknown. If
the x-ray intensity from a specimen of unknown composition is compared to that from a specially constructed
standard of very similar composition and physical
properties, then deviations from equation (1) will be
minimized. A logical flaw in this procedure stems
from the fact that it presumes from the outset that
one knows the composition of the unknown specimen. In
addition, it is dependable only insofar as the standard accurately reflects not only the true composition, but also the physical properties, of the sample.
This restriction necessitates tedious and often time-
consuming standard preparations. Usually one has little control over physical properties such as density.
Furthermore, such standards will usually not reflect
the heterogeneity inherent in biological tissue. For
instance, if the composition of a standard and a specimen were macroscopically identical, but the standard
was homogeneous, whereas the specimen contained het-
erogeneous inclusions, then x-ray analysis of a micro-
volume could clearly give disparate results.

Another technique commonly used in quantitative
microprobe analysis has been developed by Hall (1968
and 1971). This procedure utilizes a unique

Computer Correction of Intensities

theoretical approach based on the fact that the continuum radiation gives an indication of local mass per unit area. This technique has been successfully applied to a number of biological systems; nevertheless, there are requirements to the procedure which complicate its use and restrict its applicability. Of particular importance, the procedure requires that the average energy loss of the incident electrons to the specimen or standard be very small, and x-ray absorption within the sample or standard be negligible. This technique is therefore only applicable to very thin specimens and standards which are mounted on electron transparent substrates. If the sample is not "thin", if thin standards are not available, or if substrates which are electron opaque are required, then analysis using this technique is not feasible.

Due to these limitations in available biological quantitative techniques, we decided to develop a new quantitative procedure which would be both dependable and generally applicable to most biological analyses. Specifically, we wanted a technique which was
   a) accurate;
   b) simple to use — one should not have to measure any correction factors or look up parameters such as fluorescence yields;
   c) convenient — one should be able to use unmodified equipment and easily obtainable metallurgical and mineralogical standards;
   d) applicable to material of any thickness; and
   e) applicable with or without any substrate.

To a large degree, these first three requirements were already met by the existing quantitative techniques of the mineralogical and metallurgical fields. These techniques are an amalgam of basic theory and experimental determinations, as illustrated in Figure 1. This interplay between theory and experiment has led to a constant improvement in the accuracy of this type of quantitative analysis. As shown in Figure 2,

251

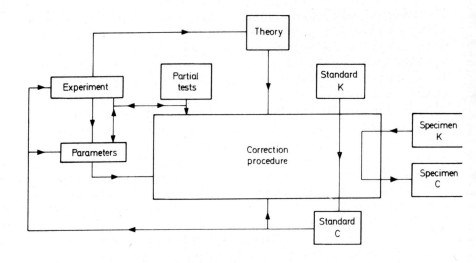

Figure 1. *Block diagram of relationships between theory and experiment involved in developing and refining correction procedures, (from Heinrich (1971)).*

most metallurgical and mineralogical analyses can at present be expected to fall within ±1% of the true value. Figure 3 is a histogram showing the distribution of errors analytically obtained in the metallurgical and mineralogical fields before the turn of the decade. Determinations which do not meet the criteria for an accurate analysis, criteria which were elegantly established by Yakowitz and Heinrich (1968), have been eliminated from this histogram. Most of the determinations are within ±1% of the true value.

The question to be asked is, can the application of metallurgical and mineralogical techniques to biological specimens result in similar accuracy. The answer is, presently, no. The results described above are usually not obtained when studying low concentrations, light elements (C, N, O), thin films, or systems where mass absorption coefficients are uncertain — precisely the conditions normally encountered in biological analyses. However, although

Computer Correction of Intensities

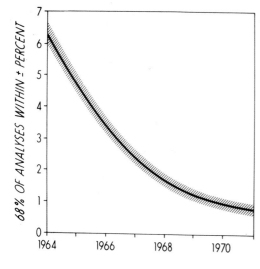

Figure 2. *Historical summary of the improvement in quantitative accuracy (from Beaman and Solosky (1972)).*

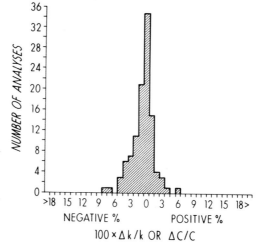

Figure 3. *Histogram of quantitative analyses performed with crystal spectrometers (from Beaman and Solosky (1972)).*

metallurgical and mineralogical quantitative techniques would not be as accurate when applied to biology, there is considerable leeway in making this technique as accurate as other biological techniques, which in general have an accuracy in the range of 10–20% (Hall, 1968). In addition, the prospects for good accuracy in applying mineralogical and metallurgical techniques to biology are perhaps greater than might at first appear. Although the major constituents of biological tissue are generally C, N, and O, the analysis of which would involve relatively large errors, these constituents are frequently of little interest to biologists. Indeed, with the other quantitative techniques one usually assumes the presence of a "typical" organic matrix. The elements which are of interest to a biologist generally have more accurately known parameters and will in general be more accurately determined. We therefore decided to assess the applicability and accuracy of metallurgical and mineralogical techniques when applied to biology.

As previously discussed, quantitative microprobe analyses are based on equation (1). A more exact expression is given by

$$C_A = \gamma_A K_A \qquad (2)$$

where γ_A is the proportionality factor correcting for the deviation between concentration and measured emergent x-ray intensities. In our correction procedures, this proportionality factor is calculated by the conventional metallurgical and mineralogical multiplicative ZAF procedure (Philibert and Tixier, 1968). As illustrated in Figure 4, the emergent x-ray intensity ratio is reduced by the fraction of measured radiation with the same wavelength which has been emitted due to secondary x-ray excitation rather than primary electron excitation (fluorescence, or F); increased by the fraction of radiation that has been absorbed in passage out of the target (absorption, or A); and corrected for the non-linearity between intensity ratio and atomic number (atomic number effect, or Z). Consequently, one can write equation (2) as

$$C_A = Z \cdot A \cdot F \, K_A \qquad (3)$$

This equation, in schematic form, is the complete correction procedure.

Computer Correction of Intensities

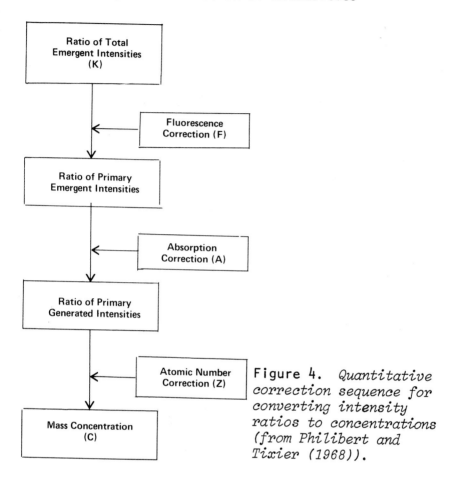

Figure 4. *Quantitative correction sequence for converting intensity ratios to concentrations (from Philibert and Tixier (1968)).*

The existing metallurgical and mineralogical techniques could not be directly applied to biological analyses but required rather extensive modifications. In addition, it was discovered that no single technique could easily encompass all of the objectives we desired for a generally applicable procedure. Consequently, three procedures, MIC, BICEP, and BASIC, have been devised and formulated as computer programs (Warner, 1972). Computers are necessary since some correction parameters are functions of concentration, which necessitates the use of successive approximations; hand computations become formidable.

THE MIC PROCEDURE

MIC, an expansion and modification of Colby's MAGIC IV (Colby, 1971), is a correction procedure specifically designed for the analysis of thick biological material. Many of the features available in MIC will be discussed in more detail with BICEP. Three salient features of most biological analyses, however, severely limit the usefulness of MIC: most biological specimens are of low density, are usually analyzed when only a few micrometers thick to improve cellular detail, and are analyzed at high accelerating voltages to detect the low concentrations of physiologically interesting elements. Under these conditions most biological specimens are thin in relation to the depth of electron penetration. Consequently the MIC quantitative procedure, although generally applicable to thick biological specimens, will have limited applicability to most biological problems.

BICEP CORRECTION PROCEDURE

BICEP, a modification of Colby's thin film model (Colby, 1968), is a correction procedure designed primarily to calculate ratios between elements present within the analyzed volume of thin biological material. Chemical symbols and the spectral lines analyzed, K-values, chemical symbol of the substrate (if any), density, specimen thickness, and accelerating voltage are the only input required for program operation. The well characterized and convenient mineralogical and metallurgical standards can be utilized.

In order for any quantitative analysis to be accurate, all elements present in the analyzed volume must be included in the correction procedure. However, the major constituents of a biological analysis, carbon, nitrogen, and oxygen, are elements which are difficult to adequately detect and measure. One of the more important features, not only of BICEP but of

256

the other computer programs as well, is the elimination of the necessity to measure intensity ratios for every element that is present. By subtracting the sum of the analyzed elements from 100%, the concentration of one element may be determined by difference. In addition, the concentration of one element may be determined by a fixed ratio to an analyzed element. And of primary importance, one can simply assume concentrations. The concentrations of these unanalyzed elements are entered into the quantitative calculations for the analyzed elements. The advantage of these capabilities over the previously available biological quantitative techniques is that one can easily model high and low values of an unanalyzed element to determine the effects of any assumptions. As illustrated in Table I, these values can then be used to set confidence limits for the analysis.

TABLE I

C	82.21*	—	—
N	—	82.13*	—
O	—	—	81.92*
Ca	4.69	4.66	4.64
P	8.06	8.04	8.06
Mg	1.88	1.96	2.08
K	1.15	1.14	1.14
Na	1.47	1.53	1.63
Cl	0.55	0.54	0.54

*Determined by difference

The effect of variation of organic material on the analyzed inorganic composition of a refractile granule from *Tetrahymena pyriformis* (values in weight percent). Values for Na and Mg differ by as much as 11%. The other constituents are relatively unaffected.

Other features are present which enhance the versatility and flexibility of BICEP. For instance, the program can be run backwards — one can input concentrations or accelerating voltages and calculate the expected intensity ratios. This ability to easily model a biological analysis can be used to determine quickly optimum analytical conditions.

The data correction procedure in BICEP is based upon the formulation for the atomic number correction, given by

$$
C_A = \frac{R_A \int_{E_C}^{E_O} Q/S_A \, dE}{R_{AB} \left[\int_{E_L}^{E_O} Q/S_{AB} \, dE + \eta_S \int_{E_{L'}}^{E_L} Q/S_{AB} \, dE \right]} \cdot K_A
\tag{4}
$$

where AB is the specimen containing element A, R is the backscatter loss factor, Q the ionization cross section, S the stopping power, η_S the backscattered electron coefficient for the substrate, if present, E the energy of the electrons, E_O the accelerating voltage, E_C the critical excitation potential, E_L the mean electron energy at the substrate-specimen interface, and $E_{L'}$ the mean energy of electrons (backscattered from a substrate) as they leave the specimen at the upper surface (= E_C if absorbed) (Colby, 1968). A physical representation of the electron trajectories involved in equation (4) is presented in Figure 5. E_L and $E_{L'}$ are calculated from the range equation of Andersen and Hasler (1966),

$$
\rho x = .032 \, A/Z \, (E_O^{1.68} - E_L^{1.68})
\tag{5}
$$

where ρ is the density, x the specimen thickness, A the average atomic weight, and Z the average atomic number. A more detailed discussion of the inner workings of BICEP is presented elsewhere (Warner and Coleman, 1973).

Computer Correction of Intensities

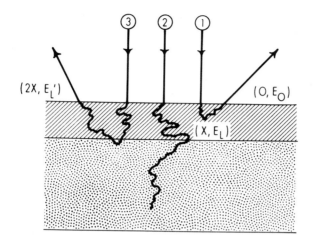

$$C_A = K_A \frac{R_A \int_{E_C}^{E_O} Q/S_A \, dE}{R_{AB} \left[\int_{E_L}^{E_O} Q/S_{AB} \, dE + \eta_S \int_{E_{L'}}^{E_L} Q/S_{AB} \, dE \right]}$$

Figure 5. *Electron paths through a thin sample mounted on an electron opaque substrate (from Warner (1972)).*

The accuracy of BICEP was assessed by the analysis of a series of aluminum films, a carbon film, and a film of a chlorinated hydrocarbon polymer, Parylene C. The results are presented in Table II. No error exceeds 9%.

In order to determine absolute concentrations, BICEP requires accurate knowledge of density and thickness in the analyzed volume. These quantities will in general not be known for biological material. However, as shown in Table III, although the BICEP determination of weight percent concentrations is quite sensitive to values of ρx, the ratios between analyzed elements are relatively insensitive to these

TABLE II

BICEP Analysis of Thin Films

	Depth (μm)	K–Ratio	BICEP Concentration (wt. %)
Aluminum	.03	.007	94.4
	.19	.049	99.4
	.34	.094	102.9
	.60	.173	104.7
	.79	.222	103.6
	1.28	.392	107.8
	1.38	.420	107.4
	1.81	.488	100.9
	1.67	.509	107.1
	2.30	.611	100.6
	2.25	.643	105.4
	2.41	.654	103.0
	3.05	.785	104.2
	3.43	.814	100.6
	3.85	.903	101.0
Carbon	.17	.069	108.9
Parylene C 25.6% Cl NaCl Standard	5.5	.299	24.7

Aluminum and carbon K-ratios with pure elemental standards. Chlorine analysis with respect to NaCl standard.

values. The primary usefulness of BICEP therefore lies in its ability to determine elemental ratios. As previously discussed, modeling various values for density and thickness can be used to calculate upper and lower limits for the analysis.

Computer Correction of Intensities

TABLE III

The Effect of Variations of Density and Thickness on
BICEP Calculated Concentrations and Ratios

	$\rho = 1.0$ $x = 0.5\mu m$	$\rho = 2.0$ $x = 0.5\mu m$	$\rho = 2.0$ $x = 1.0\mu m$
Ca	9.35	4.69	2.34
P	16.02	8.06	4.03
Mg	4.66	1.96	1.22
K	2.28	1.15	0.57
Na	3.21	1.53	0.88
Cl	1.09	0.55	0.27
O*	28.83	14.51	7.26
C**	34.56	67.55	83.43
Ca/P	0.584	0.582	0.581
Mg/P	0.291	0.243	0.303
K/P	0.142	0.143	0.141

*Determined by ratio to phosphorus
**Determined by difference

Although halving both density and thickness
nearly quadruples the absolute weight percent
concentrations, the least accurately known
ratio varies at most by 25%.

THE BASIC PROGRAM

BASIC, the last computer program we will discuss,
has been designed primarily to calculate the absolute
concentrations present in thin biological material.
In addition to concentrations, BASIC simultaneously
calculates values for specimen ρx; if specimen thickness is known, one can then calculate molarities
(grams per 1000 cm^3 dehydrated tissue). Most of the
features of BICEP are present in BASIC. The only required inputs for computer operation are the accelerating voltage, chemical symbols and spectral lines
analyzed, and the K-values.

Central to the computational process in BASIC is the use of the substrate signal generated beneath the specimen. Consequently BASIC is limited to specimens sufficiently thin to permit detection of the substrate signal. This restriction will only rarely limit the applicability of the procedure. BASIC is also restricted to specimens mounted on electron opaque elemental substrates.

The correction procedure in BASIC is similar to that in BICEP and is based on equation (4). In order to integrate this equation, the values of E_L and $E_{L'}$ must be known. These values were obtained in BICEP by using equation (5), which requires knowledge of density and thickness. To circumvent this requirement in BASIC, E_L is calculated from the substrate intensity using the power law of generated characteristic radiation. With the electron beam placed on exposed substrate, characteristic radiation is generated according to the equation

$$I = \eta(E_O - E_C)^{1.63} \qquad (6)$$

where η is the efficiency of x-ray production (Green, 1963). The substrate intensity generated beneath a specimen will be given by

$$I = \eta(E_L - E_C)^{1.63} \qquad (7)$$

From the ratio of these two equations, one can calculate the value of E_L. This procedure provides an internal compensation for variations in specimen density and thickness (Tousimis, 1971). Since the generated intensities are required by equations (6) and (7), measured intensities are corrected for absorption by an empirical correction which was determined from the analysis of aluminum films. The results from these aluminum films, as well as the carbon and Parylene C films, are presented in Table IV. No error exceeds 14%. A more detailed description of BASIC is

Computer Correction of Intensities

TABLE IV

BASIC Analysis of Thin Films

	Depth (μm)	K-Ratio	BASIC Concentration (wt. %)
Aluminum	.03	.007	86.9
	.19	.049	97.4
	.34	.094	100.4
	.60	.173	105.4
	.79	.222	102.8
	1.28	.392	112.0
	1.38	.420	109.9
	1.81	.488	98.3
	1.67	.509	109.3
	2.30	.611	98.8
	2.25	.643	105.0
	2.41	.654	102.1
	3.05	.785	103.5
	3.43	.814	97.6
	3.85	.903	97.7
Carbon	.17	.069	109.2*
Parylene C 25.6% Cl	5.5	.0299	23.6

*Analysis Conducted at 10 keV

Aluminum and carbon analysis with respect to elemental standards. Chlorine with respect to NaCl standard.

presented elsewhere (Warner and Coleman, submitted for publication).

SUMMARY

In conclusion, three quantitative procedures utilizing computer programs have been devised for the analysis of biological specimens. These programs are

based on conventional quantitative techniques of the
metallurgical and mineralogical fields. Each program
is applicable to a particular set of parameters commonly encountered in biological analyses. The programs, written in Fortran IV, are easy to use, require
no instrument modifications or special techniques, use
conventional and convenient metallurgical and mineralogical standards, and can be used to model biological
samples in order to establish optimum analytical conditions or evaluate the effect of assumptions. MIC
is applicable to the analysis of thick biological
specimens. BICEP is applicable to thin specimens
where results are best expressed as ratios, or specimens where the density and thickness are adequately
known. These two programs will be combined into one
program in the near future, thus satisfying our original objectives. BASIC is applicable to thin specimens mounted on thick elemental substrates and can
directly calculate absolute concentrations and specimen ρx.

The accuracy of these programs is comparable to
that of the previously available biological quantitative techniques. The advantages of these computer
programs are their flexibility, wide applicability,
and ease of application. In addition, the accuracy
of these programs is not limited by standards or assumptions, but by the equations of the correction procedure. The various parameters in these equations,
and indeed some of the equations themselves, are under-
going continual analysis and refinement in all the
related microprobe fields. The accuracy of these programs therefore has good prospects for continued improvement.

ACKNOWLEDGEMENTS

This paper is based on work performed partially
on NIH Grant No. AM-14272 and Biophysics Training
Grant No. 5T1GM-1088 and partially under contract with

Computer Correction of Intensities

the U. S. Atomic Energy Commission at the University of Rochester Atomic Energy Project and has been assigned Report No. UR-3490-310.

REFERENCES

Andersen, C. A., and Hasler, M. F., 1966, In, *X-Ray Optics and Microanalysis*, Castaing, Descamp and Philibert, Eds., Hermann, Paris, 310-327.

Beaman, D. R., and Solosky, L. F., 1972, *Analytical Chemistry*, 44, 1598-1610.

Castaing, R., 1951, Ph.D. Thesis, Univ. Paris, ONERA Publ. No. 55.

Colby, J. W., 1968, In, *Advances X-Ray Anal.*, Newkirk, Mallett and Pfeiffer, Eds., Plenum Press, New York, 11, 287-305.

Colby, J. W., 1971, *Proc. 6th Natl. Conf. on Electron Probe Analysis*, Pittsburgh, Penn., 17A-17B.

Green, M., 1963, In, *X-Ray Optics and X-Ray Microanalysis*, Pattee, Cosslett and Engstroem, Eds., Academic Press, New York, 185-192.

Hall, T. A., 1968, In, *Quantitative Electron Probe Microanalysis*, K. F. J. Heinrich, Ed., N.B.S. Special Publ. 298, 269-299.

Hall, T. A., 1971, In, *Physical Techniques in Biological Research*, G. Oster, Ed., Academic Press, New York, 157-275.

Heinrich, K. F. J., 1971, Lecture, Lehigh University, Bethlehem, Pennsylvania.

Philibert, J., and Tixier, R., 1968, In, *Quantitative Electron Probe Microanalysis*, K. F. J. Heinrich, Ed., N.B.S. Special Publ. 298, 13-33.

Tousimis, A. J., 1971, *Proc. 6th Natl. Conf. on Electon Probe Analysis*, Pittsburgh, Penn. 36A.

Warner, R. R., 1972, Ph.D. Thesis, Univ. of Rochester, Rochester, New York.

Warner, R. R., and Coleman, J. R., 1973, *Micron*, 4, 61-68.

Warner, R. R., and Coleman, J. R., *A biological thin specimen microprobe quantitation procedure that calculates composition and* ρx. Submitted for publication.

Yakowitz, H., and Heinrich, K. F. J., 1968, *Microchimica Acta*, 1, 182-200.

DISCUSSION

QUESTION: If I would like to measure sodium in a cell, what would I have to do to use your program? What measurements must I make?

ANSWER: The intensity ratios. You would have to measure the usual intensity ratio — that is, the X-ray count of sodium from the cell divided by the X-ray count of sodium from an appropriate standard. One should also measure intensity ratios for all elements (in addition to sodium) which are present in the analyzed volume, since these elements can affect the sodium signal. Since it is difficult to accurately measure the ratios for carbon, nitrogen, and oxygen, one can instead assume values for these elements and then assess the effects of the assumptions. Incidentally, some care should be exercised in choosing an

Computer Correction of Intensities

appropriate standard. Some are better than others, and you can find this out with the program; it prints out how good it is.

QUESTION: Does it matter whether the sample is embedded, dried, or frozen?

ANSWER: It does not matter, particularly with BASIC, which internally compensates for variations in density and thickness.

QUESTION: How does the program know what these various factors are? What do I need to feed into it?

ANSWER: This depends to some extent on which program you use. With BASIC, you would have to enter the substrate intensity ratio. With BICEP, you have to enter a value for density and thickness, although for the calculation of elemental ratios, these values are extremely uncritical. Nothing is required for MIC. In all of these programs the answer is expressed as the concentration of the element of interest in the prepared sample. If your sample is embedded, you will get the concentration of sodium in sample plus embedding medium.

QUESTION: What kind of computer do you need for BASIC?

ANSWER: We use an IBM 360. We haven't had any experience with other computers, but I know that Colby's MAGIC, from which these programs were derived, has been modified for a large number of different computers. The programs don't require too much storage space — most of the parameters are calculated internally. Basically, it has the same storage space as MAGIC; of course, one can break it down to a very small size if one is willing to do the programming.

QUESTION: Have you compared the concentration ratio from BICEP with the value you get from the scaler?

R. R. Warner and J. R. Coleman

ANSWER: Yes. For instance, from Table II, the "apparent" concentration of aluminum in a pure aluminum film .03 μm thick is 0.69%. This is the intensity ratio from the scaler with reference to an infinitely thick pure aluminum standard. The value for the aluminum concentration of this film after correction by BICEP is 94.4%.

THE DIRECT ELEMENT RATIO MODEL FOR QUANTITATIVE ANALYSIS OF THIN SECTIONS

JOHN C. RUSS
EDAX Laboratories
Raleigh, North Carolina, USA

THIN SECTIONS FOR ANALYSIS

Biological specimens can be analyzed in the form of bulk tissue, thick sections on substrates, or thin sections mounted on grids. Of these methods, the latter offers the combined advantages of the best quality image and best x-ray resolution and sensitivity, admittedly at the expense of more difficult specimen preparation. The transmitted electron image, either in a conventional TEM or scanning EM with transmission detector, shows internal structure far better than the surface image from bulk material, and has superior image resolution to the backscattered electron image from thick sections on substrates. Also, the reduction of lateral scattering of electrons in thin sections gives the possibility to analyze smaller areas, usually under 1000 Å (100 nm) diameter and as little as a few hundred angstroms in some cases. The sensitivity (minimum detectable limit) of a few hundred ppm which can usually be obtained by electron excited x-ray analysis in tissue would thus be only 10^{-18} to 10^{-19} grams of the element, but because the signal strengths from such small amounts are so low, 10^{-17} to 10^{-18} grams is a more realistic lower limit for detection of most elements.

WHAT IS "THIN"?

The term "thin sections' is generally identified with specimens prepared suitable for transmission electron microscopy. However, particularly with the advent of scanning transmission electron microscopy, these thicknesses can range from a few hundred angstroms up to a few microns. For purposes of the analytical model to be described for quantitating results, it is necessary to define "thin" as enough that the characteristic intensity for an element is linearly proportional to the mass of the element present, or to the product of the elemental concentration and specimen thickness. These are not entirely equivalent, as we shall see.

The effects to be considered are the absorption of generated x-rays in the tissue, and the slowing down of incident electrons which changes their probability for exciting atoms and also the diameter of the excited volume. The maximum x-ray path length over which absorption can be neglected can be estimated from the x-ray absorption equation

$$I/I_o = e^{-\mu \rho t}$$

If we will accept a 10% absorption, this gives a dimension t (in micrometers) of 0.1 $(\mu \rho)^{-1}$, where $(\mu \rho)^{-1}$ is the absorption mean free path in micrometers. For calcium Kα x-rays in tissue this would correspond to a thickness of more than 10 μm, while for sodium Kα x-rays, it may be only about 0.6 μm. The ten micrometer thickness, however, is certainly so thick that significant electron scattering will take place.

To estimate the thickness for which this becomes a factor, we measured the Bremsstrahlung intensity as a function of thickness from sections of epoxy resin. At an accelerating voltage of 40 kV, the curve deviates from linearity by 10% for sections approximately 0.7 μm thick. It was noted, however, that the same

270

intensity was measured above and below the specimens for specimens in excess of 1 μm thick, suggesting that absorption is not significant at this thickness. The deviation is thus due to scattering of the electrons, and with their change in energy, a change in the ionization cross-section. From these limited measurements, we expect that for purposes of our analytical model, an upper limit of applicable section thickness is in excess of 0.5 μm. The use of ratios of intensities probably extends this somewhat and accuracy for sections up to at least 1 μm is actually observed.

RELATING INTENSITY TO CONCENTRATION

In the model proposed by Hall, and rather widely used for quantitative analysis, the ratio of net peak intensity to Bremsstrahlung intensity is proportional to concentration. This arises from the linearity of characteristic intensity to elemental mass, and the linearity of Bremsstrahlung intensity with total excited mass, both of which follow from our "thin-ness" assumption. Using the Hall model requires the use of standards, either in the form of similar sections of tissue-like material with dissolved concentrations of elements, or high concentration sections such as minerals from which calculation of the ratios can be made.

However, the determination of absolute concentrations of elements in tissue is of doubtful validity, because of the addition or removal of mass in preparation and the loss of organic mass under the electron beam. For these reasons, many workers prefer instead to report elemental concentrations in terms of ratios such as Na/K or Ca/P. Using the Hall model, this would give for a specimen

$$\frac{(I/B)_1}{(I/B)_2} = K_{12} \frac{C_1}{C_2}$$

or, since the background will cancel, simply

$$C_1{:}C_2{:}C_3{:}\ \cdots\ =\frac{I_1}{P_1}:\frac{I_2}{P_2}:\frac{I_3}{P_3}:\ \cdots\ ,$$

for each of the elements in the analyzed volume. The P values are proportionality constants that could be determined from standards. In fact, Duncumb has used this method for metal thin sections by using the over-all specimen analysis obtained with a spread beam, and the known average composition, to determine the pro-portionality factors.

It is possible, however, to eliminate the need for standards by calculating the P values directly, since it is simply the product of the probabilities that:

      a) the incident electrons ionize an atom of the element;

      b) the atom emits an x-ray of the line being analyzed; and

      c) the x-ray is detected.

This can be written as

$$P = Q \cdot W \cdot R \cdot T$$

where Q is the ionization cross-section, W is the fluorescent yield, R is the relative intensity of the line being analyzed and T is the spectrometer effi-ciency. The first three terms are readily calculated or found in tables. The final term can be determined experimentally for a given x-ray spectrometer. For an energy-dispersive spectrometer, it is convenient to express it as

$$T = \delta \cdot e^{-0.58/E^3}(1 - e^{-18700/E^3})$$

where δ is the solid angle, E is the x-ray energy, and the constants describe a system with a 7.5 μm thick Be window and a 3 mm thick active volume.

These values of P will convert net intensity values I to relative atomic fractions. To obtain

Direct Element Ratio Model

weight fractions, the atomic weight A can be added.

$$P = Q \cdot W \cdot R \cdot T/A$$

The expression can be straightforwardly evaluated for the accelerating voltage(s) commonly used, and a table of P values prepared.

APPLICATION AND TEST OF THE MODEL

The direct calculation of elemental ratios using this simple model was tested by analyzing particles of known stoichiometric particles dispersed on a carbon-coated formvar film. The results for particles of various sizes (in the projected image) are shown in Table I.

The overall accuracy appears to be about 10% relative. The fact that acceptable results are obtained for particle sizes much greater than the upper limit suggested for good thin section analysis is considered to be partly due to the fact that the particles may be smaller in thickness than projected size. In addition, and probably more important, the use of intensity ratios makes the method less sensitive to thickness since the characteristic intensities from the various elements are affected more or less similarly by increasing thickness.

In a second test, the analyses of fragments of intermetallic compounds from an iron-zinc alloy (galvanized coating on steel) was obtained as shown below. The concentrations correspond closely to the expected intermetallic compounds $Fe\,Zn_7$ and $Fe\,Zn_9$.

AREA	ELEM.	LINE	INT.	AT. %	WT. %
1	FE	KA	13472	12.45	10.83
	ZN	KA	75115	87.55	89.17
2	FE	KA	11662	9.92	8.60
	ZN	KA	84041	90.08	91.40

TABLE I

Particle	Size (diameter)	Intensities	Atomic ratios calculated from the intensities
Manganese sulfate	12.0 µm	16917 : 16074	1 : 0.93
	2.5	1435 : 1393	1 : 0.95
Mn : S	1.1	1282 : 1349	1 : 1.03
1 : 1	0.8	497 : 467	1 : 0.92
	0.4	336 : 303	1 : 0.88
Calcium alumino-	5.5	4788 : 6839 : 7932	1 : 2.02 : 2.05
silicate	1.2	5245 : 8085 : 8180	1 : 2.18 : 1.93
	0.6	633 : 1008 : 997	1 : 2.15 : 1.86
Ca : Al : Si			
1 : 2 : 2			
Iron chloride	11.5	38440 : 115674	1 : 2.73
Fe : Cl	2.3	3078 : 10280	1 : 3.03
1 : 3	0.9	1246 : 4435	1 : 3.23
	0.7	960 : 3345	1 : 3.16

Direct Element Ratio Model

This method has also been used for quantitative analysis of elemental ratios in biological tissue. The data shown below were obtained from Golgi apparatus in a section of kidney from a guinea pig. The osmium is from the fixative and the chlorine from the epoxy resin. The accuracy of the results is believed to be better than 10% relative, but because of the effects of specimen preparation, the concentrations are not believed to be representative of the living condition.

KV = 27.2

ELEM.	LINE	INT.	ATOM. RATIO
CA	KA	1242	1.000
K	KA	812	.66
S	KA	356	.33
CL	KA	599	.52
NA	KA	308	.67
MG	KA	74	.116
OS	MA	1028	2.24

To convert these relative concentrations to absolute values, it is only necessary to determine any one element. This can be done using the Hall model with standards, or by the inclusion of an internal standard. We have found the Cl present in the embedding media to be suitable for the latter role in some cases.

CONCLUSION

The direct element ratio model described is used in the analysis of thin sections to convert net intensities for the elements present to relative concentrations. It is estimated to give 10% or better relative accuracy in the results, and to be applicable to sections ranging up to 0.5 to 1.0 μm in thickness. It is not sensitive to specimen preparation, provided, of course, that the elements of interest are not moved or removed. The method has the virtue of extreme

simplicity, and yet gives accuracy at least as good
as other more complex models. The estimate of the
maximum specimen thickness is conservative, especially if the very light elements (Mg and Na, for example)
are not being analyzed. The model can be used with
either TEM, SEM or microprobe, and is applicable to
inorganic materials as well as tissue sections.

DISCUSSION

KOENIG (Battelle-Frankfurt) showed the possibility of
introducing terms for electron scattering and absorption into the simple model described. Then with appropriate assumptions and integration, the model can
be extended to a finite section thickness.

HALL (Cambridge University) pointed out that the use
of elemental ratios extends the effective linearity of
the results beyond the specimen thickness that would
satisfy the thin-ness criteria for a single element
intensity.

X-RAY SPECTROSCOPY IN THE SCANNING ELECTRON MICROSCOPE STUDY OF CELL AND TISSUE CULTURE MATERIAL

G. M. HODGES* AND M. D. MUIR**
*Department of Cell Pathology, Imperial
Cancer Research Fund, London, England
**Department of Geology, Royal School
of Mines, London, England

INTRODUCTION

In the study of specific structures where chemical or biochemical analyses are required, it is frequently difficult to obtain sufficient and pure samples of the particular structure under study in order to carry out accurate chemical analysis by conventional means. Methods which will permit the correlation of elemental distribution with morphology are therefore of considerable value. Thus, X-ray spectroscopy using the electron probe microanalyser or the SEM can allow chemical characterisation to be correlated with structural topography (Castaing et al., 1966; Tousimis, 1968; Robertson, 1968; Birks, 1971; Ingram et al., 1972; Salisbury et al., 1972) and should be of value in analytical studies of such structures as the basement membrane which cannot, in general, be easily separated from the adjoining tissues.

One way in which this problem of chemical characterisation can be approached is by the use of heavy-metal labelled macromolecules, either in in vivo or in vitro incorporation studies, or as a staining procedure for tissue sections (Clarke et al., 1970; Rosen, 1972; Salisbury et al., 1972). There are several

factors which may affect the success of this approach
(Oster, 1971; Lifshin and Ciccarelli, 1973) and the
validity of the results is limited by the degree to
which they can be evaluated at the present time.
Among the various factors can be listed more specifically:

a) the choice of label (preferably of high
atomic number) for optimal elemental analysis,

b) the molecule size and number of bound
molecules for adequate spectral resolution,

c) the possible modification of macromolecular properties following heavy-metal labelling,

d) the possible breakdown and loss of metal
label when exposed to the electron beam.

As part of a programme on the study of basement
membranes, a pilot project was initiated to ascertain
the potentialities and limitations of a SEM X-ray
microanalytical approach. In this preliminary study,
an attempt has been made to determine whether qualitative information on the presence of carbohydrate
residues on the cell surface could be obtained by SEM
X-ray spectroscopy of glutaraldehyde-fixed monolayer
cell cultures and bladder tissue sections, following
"staining" with a heavy-metal labelled macromolecule
mercury-labelled concanavalin A.

MATERIAL AND METHODS

General

Bladders, dissected from C57at Icrf 6 months-old
mice, were placed in Hanks saline solution and 2 —
3 mm^2 explants prepared. One group of explants was
placed in 3% trypsin — 1% pancreatin solution for 1½ h
at 37°C, while a control group of explants was placed
in Hanks saline solution. At the end of the incubation period, the tissues were fixed in 2.5% glutaraldehyde in 0.1 M sodium cacodylate buffer pH 7.1 for
24 h at 4°C, washed twice in buffer over a period of

Cell and Tissue Culture Material

48 h at 4°C, and embedded in paraffin wax. Sections
5 μm thick were cut and mounted on 13 mm coverslips.

HeLa cells grown to confluency on 13 mm glass
coverslips were treated either with 0.1% trypsin or
Hanks saline solution for 5 minutes at 37°C, fixed in
2.5% glutaraldehyde 0.1 M cacodylate buffer for 24 h
at 4°C, rinsed twice in buffer for 48 h at 4°C, washed
twice in distilled water and rapidly air-dried.

Mercury-labelled Concanavalin A

This material prepared according to the method
of Horisberger *et al.*, (1971), was kindly provided by
Dr. C. Rowlett. 10 mg of Hg-labelled con A was dissolved in 10 ml 0.06 M Sorensen's phosphate buffer
containing 1 M NaCl and 0.001 M $MnCl_2$ and the solution
filtered through a 0.45 μm pore-size Millipore membrane filter.

Treatment of Cells and Tissues with Hg-labelled Concanavalin A

The Hg-labelled con A solution was added to control and enzyme treated HeLa cells and to bladder tissue sections for 2 h and maintained at 25°C. The
specimen was then washed twice in Sorensen's buffer,
twice in double distilled water, rapidly air-dried,
and mounted on Dural specimen stubs. Uncoated and
gold-coated specimens were examined in a scanning
electron microscope (Cambridge Scientific Instruments
Co. "Stereoscan" Mk IIA)

Equipment Used in the X-ray Investigation

The SEM is equipped with a fixed position Ortec
Si(Li) solid state energy dispersive detector which
is coupled with a Northern Scientific Econ II multi-
channel analyser. We are grateful to Mr. D. A. Lock
of Tracor (UK) Ltd. who lent us a variable geometry
detector for comparison purposes. Both detectors have

thin windows, the Ortec window being 10 μm thick, while the Tracor detector has a window 7.5 μm thick. In addition, the Tracor detector can be moved over a range of about 3.5 cm from the normal position, where it is flush with the inner wall of the SEM specimen chamber, to a position where the window of the detector is within 5 mm of the specimen surface. No accurate measurements of the distance detector/specimen surface are available.

Some tests were carried out to determine the effects of moving the detector from the distant (out) position to the near (in) position. Figure 1 shows the two signals in a log display. The upper spectrum shows the output produced when the detector is in the "in" position, while the lower spectrum shows the output when the detector is in the "out" position. It is clear that both spectra are essentially the same, since all the peaks, with one exception, are represented in the two spectra. The one exception, shown in greater detail in Figure 2, is the Fe Kα peak which is always present in our X-ray spectra and is the result of back-scattered primary electrons exciting Fe X-rays from the bottom plate of the final lens assembly of the Stereoscan. It is difficult to exclude this spurious information by means of collimation of the detector when the detector is in the "out" position, but the use of a variable geometry detector can eliminate this particular problem. This is of importance in the determination of Fe in biological specimens.

Use of the variable geometry detector is valuable also in that since the solid angle of X-rays collected is much greater when the detector is in the "in" position, count rates are considerably improved. This point is illustrated in Figure 3, which shows the initial part of the spectrum shown in Figure 1, but displayed this time on a linear scale. It is clear that the number of counts in the Cl Kα peak obtained in the "in" position is considerably higher than the number

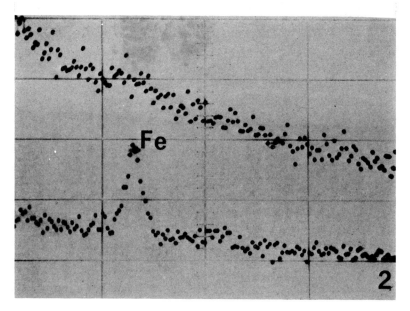

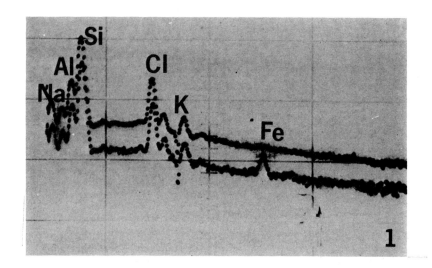

Figures 1 and 2. *Comparison of emission spectra (log display) from bladder section using a retractable solid state Si(Li) detector; bottom spectrum in a distant (out) position, top spectrum in a near (in) position. Note absence of Fe Kα peak in the "in" position.*

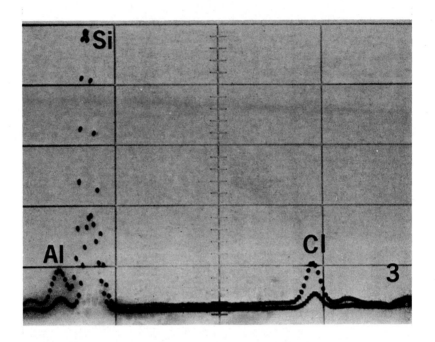

Figure 3. *Comparison of emission spectra (linear display) from bladder section showing the increased count rates obtained when the detector is in the "in" position.*

of counts in the "out" position. The same was true for the K Kα peak. The results of a comparative analysis of a mineral standard (chalcopyrite) using the variable geometry detector are shown in Table I, and these confirm the findings illustrated in Figure 3. However, there is a considerable drop in peak-to-background ratio resulting in a loss in detector resolution. Comparative analysis of bladder sections gives similar results in that for a number of elements, the counts obtained using the detector in the "in" position are about two and a half times the numbers of counts obtained in the "out" position (see Table II).

The analyses were in general carried out at an accelerating potential of 20 kV, beam current of 175 µA,

Cell and Tissue Culture Material

TABLE I.

Comparison of count rates for a chalcopyrite (FeCuS mineral) using a retractable solid state Si(Li) detector in a distant (out) position and a near (in) position.

Element	Peak	Back-ground	Peak-Bkgd.	Peak Bkgd.	Peak-Bkgd. Background
Detector out					
S Kα	8569	306	8263	28.0	27.0
Fe Kα	2570	85	2485	27.24	26.24
	Resolution of detector = 165 eV				
Detector in*					
S Kα	11927	2750	9177	4.34	3.34
Fe Kα	3776	662	3114	5.7	4.7
	Resolution of detector = 190 eV				

*The improvement in count rate from this mineral specimen is not so great as that from the biological tissues (see Table II). This is due to the large dead-time of the system at extremely high count rates. The high count rates also lead to pulse pile-up, which degrades the resolution of the detector and the peak-to-background ratios; these effects can be reduced by using a suitable pulse pile-up rejection system.

TABLE II.

Comparison of count rates from a bladder section (trypsin-treated, Hg-con A labelled) using a variable geometry Si(Li) detector in the "in" and "out" positions.

Element	Detector "in"	Detector "out"	Peak ratios
Al Kα	2813	1231	2.28
Si Kα	28476	11746	2.42
P Kα	1007	416	2.42
S Kα	1294	554	2.33
Cl Kα	1299	541	2.40
K Kα	2585	1114	2.32
Ca Kα	1624	729	2.21
Fe Kα	234	439	0.51
Hg Mα	841	336	2.47

and a filament current of 2.4 A. The top lens was set at 0.46 A, the middle lens at 0.48 A and the final lens at 0.66 A. The incident beam current was in the range of $2 - 8 \times 10^{-10}$A. When the detector was in the "in" position, the count rate was about 800 cps, and when in the "out" position, about 250 cps. The HeLa cells were also examined using an accelerating potential of 5 kV.

RESULTS

X-ray elemental analysis for mercury was under-taken over areas of the epithelium and of the stroma in the bladder specimens (Figure 4) and over the cell surface of the HeLa cells. The results showed that mercury could be detected in these specimens both in the control (Figure 5) and trypsin-treated material

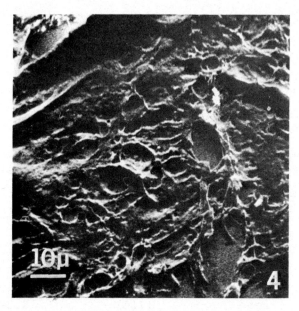

Figure 4. *Scanning electron micrograph of muscle region of uncoated, deparaffinised Hg-con A-labelled bladder section.*

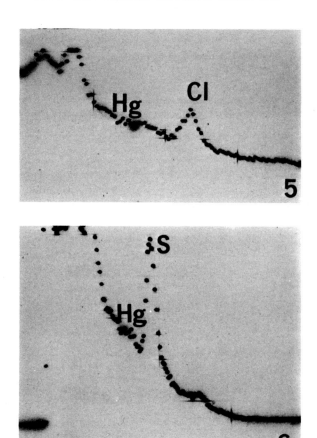

Figures 5 and 6. *Emission spectra from Hg-con A-labelled bladder sections, control (Figure 5) and trypsin-treated (Figure 6), showing low mercury concentration.*

(see Table III and Figure 6). Attempts were made to obtain a distribution pattern of the Hg X-rays, but because of the low Hg concentration, the background noise obscured any meaningful display. Furthermore, coating of the specimens with a metal conducting layer was found to reduce the overall count rates.

TABLE III.

Comparison of count rates from epithelial and muscle regions of Hg-con A-labelled bladder sections (control and enzyme-treated).

Sample	Peak	Background	Peak-Bkgd.	Peak Bkgd.	Peak-Bkgd Background
F110					
Muscle					
Na Kα	2328	1474	854	1.6	0.57
Si Kα	2591	1523	1068	1.7	0.7
Cl Kα	909	211	698	4.3	3.3
Hg Mα	644	530	114	1.21	0.21
F111					
Muscle					
Trypsin-treated					
Na Kα	1683	1490	193	1.1	0.13
Si Kα	1775	1314	461	1.3	0.35
Cl Kα	929	488	441	1.9	0.90
Hg Mα	708	494	214	1.4	0.43
F110					
Epithelium					
Na Kα	5690	3684	2006	1.5	0.54
Si Kα	4123	3924	199	1.0	0.05
Cl Kα	3368	1321	2047	2.5	1.5
Hg Mα	1741	1661	80	1.05	0.05
F111					
Epithelium					
Trypsin-treated					
Na Kα	2288	1419	869	1.6	0.61
Si Kα	2419	1749	670	1.4	0.39
Cl Kα	747	455	292	1.6	0.64
Hg Mα	694	636	58	1.09	0.09

Cell and Tissue Culture Material

TABLE IV.

Comparison of count rates from paraffin and deparaffinised sections of skin biopsy containing silver.

Sample	Peak	Background	Peak-Bkgd.	Peak/Bkgd.	Peak-Bkgd./Background
Deparaffinised section					
Na Kα	4172	1377	2795	3.0	2.0
Al Kα	5043	1377	3666	3.7	2.7
Si Kα	33628	4890	28738	6.8	5.8
S Kα	1217	1045	172	1.1	0.15
Cl Kα	1253	1037	216	1.2	0.20
Ag Lα	1055	986	69	1.07	0.07
K Kα	3063	1183	1880	2.5	1.5
Ca Kα	2206	1196	1010	1.8	0.8
Fe Kα	807	355	452	2.2	1.2
Paraffin section					
Na Kα	2122	1496	626	1.4	0.43
Al Kα	4041	2918	1123	1.4	0.38
Si Kα	4238	2531	1307	1.4	0.44
S Kα	2219	1892	327	1.2	0.17
Cl Kα	2378	1967	411	1.2	0.20
Ag Lα	1616	1534	92	1.06	0.06
K Kα	1661	1470	191	1.1	0.13
Ca Kα	1447	1192	255	1.2	0.22
Fe Kα	993	510	483	1.9	0.94

A comparison of paraffin and deparaffinised tissue sections, mounted on glass coverslips, showed that beam penetration was greater in the deparaffinised material (see in Table IV). Si count rates were found to be greatly increased while Al and Na counts showed some increase also, indicating that as a result of the loss of mass following deparaffinisation there is increased X-ray output from the glass substrate and aluminium specimen holder.

DISCUSSION

While these results demonstrate the presence of Hg in the different specimens examined and would suggest the feasibility of using heavy-metal labelled macromolecules as a "staining" procedure, a number of questions are raised which must be answered before further progress can be made.

The first point concerns the choice of label which should be selected to permit optimal elemental analysis. Elements need to be selected which (a) are preferably of high atomic number to allow for easier X-ray identification and which on spectral analysis can be easily distinguished from adjacent naturally-occurring elements, and (b) will combine in a sufficient molecular concentration with given macromolecules to form stable metallo-organic complexes. In studies of the type described in this paper where heavy-metal labelled macromolecules are used in incorporation or staining procedures it is obviously of importance to select elements of high atomic number that will be present in sufficient quantity in order to allow adequate spatial resolution in X-ray distribution studies. Another factor which must be considered in the development of such techniques is the possible modification of macromolecular properties following the incorporation of a heavy-metal label into a given macromolecule.

In the present study, possible objections may be raised to the choice of mercury which is highly volatile under vacuum and under the electron beam. However, the results show that it can be detected even at low concentrations and this does suggest that while there may be some breakdown and loss of metal label, a substance as highly volatile as Hg can be used. This leads to the important consideration of possible errors which can result from the instability of heavy-metal complexed macromolecules under electron beam irradiation and this point remains to be conclusively elucidated in the present study.

Furthermore, biological materials can suffer an intrinsic mass loss under electron beam irradiation, particularly if the SEM is operated under electron probe conditions with high (in the order of 10^{-7}A) specimen current. Many elements, for example potassium, are very volatile under such operating conditions even when complexed with biological molecules. Breakdown and loss of naturally-occurring elements and of heavy-metal labels may therefore take place and not necessarily in a uniform manner. This leads to two possible errors of interpretation; firstly, the material within the analysis area will be physically altering throughout the time taken to accumulate sufficient counts for a quantitative determination, and secondly, distribution micrographs may show areas apparently deficient in the element of interest and which are depleted by reason of this beam interaction effect. One way in which this problem may be overcome is by the use of an energy dispersive spectrometer operated at low beam currents (10^{-10} A), which minimise mass loss effects. However, in order to accumulate sufficient counts for quantitative analysis, one may need very long counting times which tend to increase mass loss effects and partially cancel the benefits achieved by using low beam currents.

Another way of minimising mass loss effects is by shock freezing the specimen and examining it at liquid nitrogen temperatures (see Gullasch and Kaufmann, this Symposium). This procedure appears to have great potential value and furthermore has the advantage of preventing migration of labile ions and molecules. Tissue preparation methods which do not involve the use of aqueous solutions for any step of the preparation are essential where distribution and quantification of soluble ions are to be assessed. Several speakers at this Symposium, (Echlin, Gullasch and Kaufmann, Hutchinson) have strongly indicated that normal methods of fixation are not very suitable for

X-ray microanalysis studies. However, observations from the present study suggest that where use of substances such as heavy-metal complexed macromolecules is involved, qualitative elemental information can be obtained using simple histological preparation methods.

In using a SEM with a solid state (Si(Li)) detector, careful consideration must be given both to the choice of specimen support and coating material (see Robertson, 1968). These must, to a certain extent, be considered in relation to the element to be analysed. In general, it is necessary to consider mounting specimens on materials of low atomic number (that is, less than the lower limit of detection which at present is fluorine Z=9) in order to minimise interference with spectral lines from contaminating elements which can occur when such substrates as glass are used (see Results section). Materials such as beryllium, carbon or certain plastics, would be suitable for the mounting of specimens and, in cases where analysis of only one element is being carried out, aluminium or SiO_2 could be usefully considered providing that their X-ray emission does not interfere with that of the element under study. Coating of specimens is found in general to lead to a reduction of counts (see Results section) and it is therefore preferable to carry out elemental analysis on uncoated material. However, where conductivity problems are present this will obviously not be possible and carbon coating would seem to be the most appropriate under these circumstances.

ACKNOWLEDGEMENTS

We wish to thank Mr. S. Fisher who prepared the Hg-labelled concanavalin A, and Mr. P. R. Grant who helped with the analyses. We are also grateful to Mr. D. Lock for lending us the variable geometry detector, and to Dr. D. A. Gedcke for the comment on

Cell and Tissue Culture Material

the figures in Table I.

REFERENCES

Birks, L. S., 1971, *Electron Probe Microanalysis*. Wiley Interscience, New York.

Castaing, R., Deschamps, P., and Philibert, J., (Eds.) 1966, *X-ray Optics and Microanalysis*, Hermann, Paris.

Clarke, J. A., Salsbury, A. J., and Willoughby, D. A., 1970, *Nature*, 227, 69-71.

Hall, T. A., 1971, In *Physical Techniques in Biological Research*, G. Oster, Ed., 1A, 2nd. Ed., Academic Press, New York. 158-275.

Horisberger, M., Baver, H., and Bush, D. A., 1971, *FEBS Letters*, 18, 311-314.

Ingram, F. D., Ingram, M. J., and Hogben, C. A., 1972, *J. Histochem. Cytochem.*, 20, 716-722.

Lifshin, E., and Ciccarelli, M. F., 1973, *Proc. 6th Ann. SEM Symposium, IITRI*, Chicago, 89-96.

Oster, G., (Ed.), 1971, *Physical Techniques in Biological Research*, 2nd. Ed., Academic Press, New York.

Robertson, A. J., 1968, *Phys. Med. Biol.*, 13, 505-522.

Rosen, S., 1972, *J. Histochem. Cytochem.*, 20, 548-551.

Salsbury, A. J., and Clarke, J. A., 1972, *Proc. R. Soc. Med.*, 65. 826-827.

Tousimis, A. J., (Ed.), 1968, *Electron Probe Microanalysis*, Academic Press, New York.

APPLICATION OF MICROPROBE ANALYSIS TO INTESTINAL CALCIUM TRANSPORT

R. R. WARNER* and J. R. COLEMAN
Department of Radiation Biology and Biophysics
School of Medicine and Dentistry
University of Rochester, Rochester, New York

INTRODUCTION

In the intestine, calcium is transported from the lumen to the circulatory system across the cells of the mucosal epithelium. The bulk of information concerning this transport process has been obtained from the measurement of calcium fluxes in whole tissue preparations, and from the detection of changes in tissue homogenates. Although these techniques have elucidated many of the parameters and characteristics of the transport process, they cannot directly resolve the activities of individual cells, and can only describe the average actions of several tissue layers and cell types. Mechanisms of calcium transport at the cellular level are therefore only poorly understood. The advent of the electron microprobe provides a new tool for the study of epithelial transport, a tool which promises to be of great use in determining precisely that information difficult to extract from conventional studies — the resolution of the activities of individual cells. The correlation of data from conventional and microprobe studies promises to accelerate the understanding of cellular transport processes. This paper on the cellular mechanism of intestinal

* Present Address: Department of Physiology, School of Medicine, Yale University, New Haven, Conn.

calcium absorption is presented as an example of the complementary interplay between conventional epithelial transport studies and microprobe analysis.

Calcium is believed to be transported across the mucosal epithelium by the so-called "absorptive cells", which together with goblet cells are the major constituents of this epithelial layer lining the intestinal tract. The transport process is believed to proceed by both active and passive mechanisms, both of which require vitamin D for optimum calcium absorption (Wasserman 1968; Wasserman and Taylor 1969). The active process is believed to be carrier mediated: calcium uptake can be saturated and obeys Michaelis-Menten kinetics (Martin and DeLuca 1969). In addition, the effect of vitamin D is mediated through protein synthesis (Harrison and Harrison 1966). These data suggest a vitamin D-sensitive protein carrier is involved in the absorption of calcium. Strong support for a carrier mechanism has been provided by isolation of a calcium-binding protein (CaBP) from intestinal homogenates (Wasserman and Taylor 1966). The physiological importance of CaBP is based on the fact that whenever or wherever calcium transport in the intestine is modified, the level of CaBP is changed in the same direction. The level of CaBP is strongly linked to vitamin D; in the absence of vitamin D, there is no CaBP. And of particular importance, the association constant of CaBP for calcium is such that even at a calcium concentration of 10^{-6} M, CaBP is half-saturated.

Any hypothesis about the mechanism of calcium transport must take into account the fact that calcium is potentially toxic to cells; for example, levels of intracellular calcium on the order of 10^{-4} M can inhibit important glycolytic enzymes (Kachmar and Boyer 1953; Kimmich and Rasmussen 1969), and uncouple oxidative phosphorylation (Chance 1963). Thus, although the extracellular medium has a calcium concentration of 10^{-3} M, it is generally believed that the concentration of calcium ion free in the cellular cytoplasm is rigorously maintained at a low level near 10^{-6} M (Borle 1967). In addition to maintaining this internal homeostasis, intestinal cells must meet the body requirement for calcium by transporting calcium from the lumen to the lamina propria. The question to be considered here concerns the cellular mechanism by which intestinal cells transport bulk quantities of calcium without raising their own cytoplasmic ion concentration.

Based on the above considerations, the following simple model for the cellular mechanism of calcium transport can be presented (Figure 1). In this model,

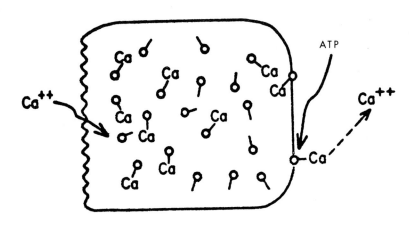

Figure 1. *A hypothetical model for the cellular mechanism of calcium transport. The lollipops indicate intracellular CaBP. See text for details.*

calcium enters the cell down its electrochemical gradient, is bound to intracellular CaBP thereby maintaining a low intracellular ionic concentration, diffuses across the cell in a bound form, and is actively pumped out at the serosal border. One can see that this model does illustrate a means of transporting calcium while maintaining a low calcium activity. In a test of this model, an immunofluorescent antibody technique was used to localize CaBP at the cellular level (Taylor and Wasserman 1970). Although the resolution of this technique was somewhat limiting, CaBP was clearly not within the absorptive epithelial cells, but was instead associated with the brush border glycocalyx of these cells and localized within goblet cells (described in Figure 2). Since CaBP appears to be present only in the brush border region of the absorptive cell, the question of how these cells translocate this potentially toxic ion remains unanswered. In addition, the presence of CaBP in goblet cells raises the question of whether these cells might also be involved in calcium transport; at present goblet cells are believed to function only in the secretion of a mucus with protective and lubricative functions.

METHODS AND RESULTS

It was felt that observing the form and path of calcium movement across individual epithelial cells would provide valuable information on the nature of the cellular calcium transport mechanism. Accordingly, experiments were conducted to determine the normal calcium distribution in this tissue, the changes in this distribution under *in vitro* conditions which promote calcium transport, and the effect of vitamin D on these distributions. Tubular segments of intestinal tissue were rapidly removed from decapitated animals and cut longitudinally to form a sheet. Tissue was then either fixed directly or incubated in a 2 mM calcium medium for various times before fixation. Tissue was cut into 1 mm^3 cubes and fixed in 6 per cent acrolein in 0.1 M cacodylate buffer, pH 7.4, containing 1 per cent sodium oxalate. Specimens were rinsed and post-fixed in 1 per cent osmium tetroxide using the same oxalate-containing buffer. Dehydration was in oxalate-saturated ethanol, followed by propylene

oxide, Epon embedding was used. This technique, described in detail elsewhere (Coleman and Terepka 1972a;
1972b; Warner and Coleman, submitted for publication;
Warner 1972), has been shown to minimize calcium loss
and redistribution, presumably through the formation
of insoluble calcium oxalate. The calcium distribution observed in tissue processed immediately after
being removed from the animal should closely reflect
the distribution occurring *in vivo*. Figure 2a shows
such an "*in vivo*" sample current image of rat intestine with the corresponding calcium Kα X-ray image. As
shown in this figure, and as seen throughout the small
intestine, calcium is associated only with areas of
mucin production and accumulation within goblet cells.
In the absorptive cells no calcium distribution above
background can be detected. Much of the goblet cell
calcium is in discrete localizations as can be seen
in more detail in figure 2b.

Goblet cells were at times seen extruding their
accumulated mucin into the lumen between villi, as
shown in Figure 3a. The calcium localizations which
had been within these goblet cells remained associated
with the extruded mucus. This latter distribution is
illustrated in Figure 3b; calcium is located extra-
cellularly in association with extruded mucus. The
mucus lies over the brush border and extends into
spaces between villi. Again it can be seen in these
figures that no calcium distribution above background
can be detected within the absorptive cells.

The pattern of calcium localizations described
above is seen throughout the small intestine. Calcium
is found associated with intracellular and extracellular goblet cell mucus. Usually no calcium is found
within the absorptive epithelial cells. On rare oc-
casions, however, in animals fed *ad libitum*, discrete
calcium localizations similar to those occurring in
goblet cells were observed within absorptive cells.
In spite of numerous line scans and point analyses
conducted on these and other absorptive cells, no

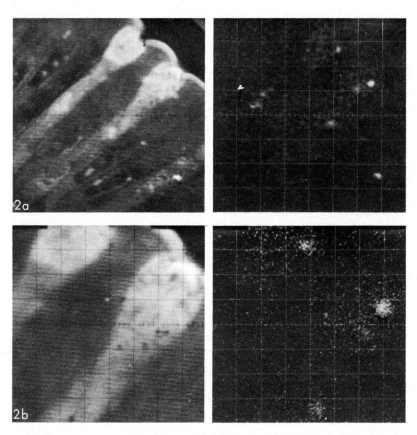

Figure 2a. *Sample current image of the epithelial layer lining the intestine. The black band in
the upper right is the intestinal lumen. The long
white cells with swollen luminal ends are the goblet
cells. The white layer bordering the lumen is the
brush border. The remaining epithelial cells with
prominent nuclei are the absorptive cells. The associated calcium Kα X-ray image shows calcium localized
only within the mucus-rich areas of goblet cells. Rat
duodenum, 6 μm/grid square.* **2b.** *Magnification of
figure 2a. Calcium is only present within goblet
cells in the region of mucin accumulation. Some of
the calcium is discretely localized, reminiscent of
the size and shape of mucigen droplets. 2.5 μm/grid sq.*

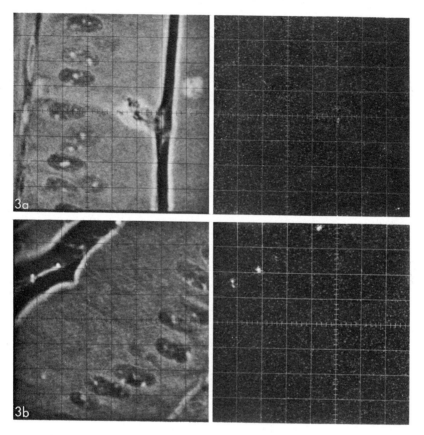

Figure 3a. *Goblet cell extruding its accumulated mucin. Although barely perceptible in this figure, some calcium is associated with the mucus which is being extruded. Rat duodenum, 9 µm/grid square.*
3b. Calcium is associated with extracellular goblet cell mucus. There is no calcium distribution above background within the absorptive cells. Rat duodenum. 8 µm/grid square.

continuous calcium concentration gradient could be observed. When calcium was detected, it was not in the form of a gradient but appeared as discrete localizations.

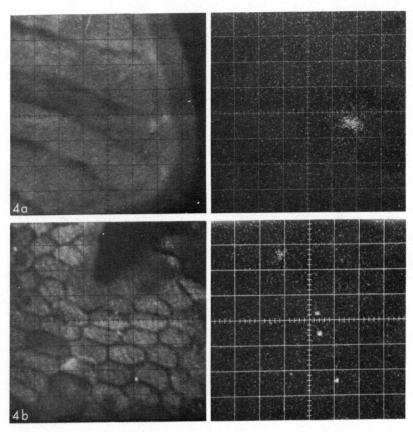

Figure 4a. *After 5 minutes of incubation, calcium is in a discrete localization near the junctional complex. Chick duodenum. 2 μm/grid square.* 4b. *A grazing section at the apical surface perpendicular to the epithelial cells, which appear to be roughly hexagonal in this plane. Fuzzy area at tissue edge is brush border. After 5 minutes of incubation, calcium localizations appear to be associated with the intercellular space near the junctional complex. Chick duodenum. 7 μm/grid square.*

Having observed the normal *in vivo* calcium distribution, incubation experiments were performed under

300

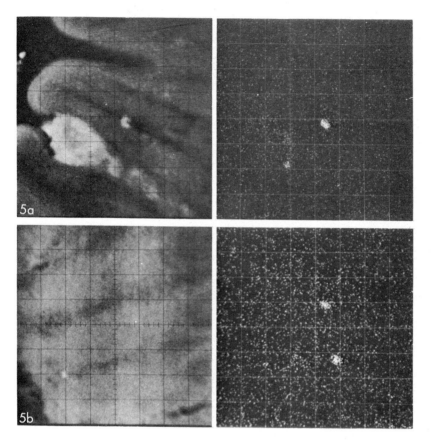

Figure 5a. *Calcium localization in the area of the lateral cell membrane; some calcium also associated with mucus in the goblet cell. Five minute incubation of rat duodenum. 4 μm/grid square. 5b. Calcium localizations below nuclei (darker areas in upper right) and appearing to be within the intercellular spaces. Cells terminate on lamina propria at lower left. 15 minute incubation of chick duodenum. 2μm/grid square.*

controlled conditions in order to determine the form and path of calcium translocation. Microprobe analysis of tissue which had been incubated for five

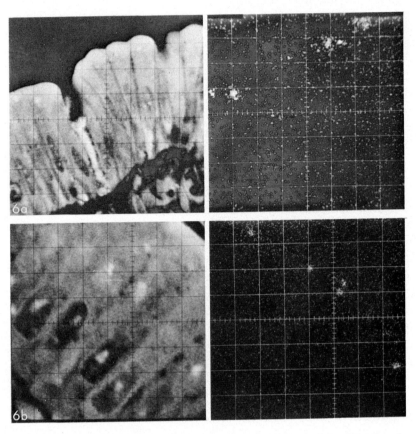

Figure 6a. *Five minute incubation of rat duodenum showing calcium localization near the apical border. The localizations at the far right and far left are apparently within the cell cytoplasm separate from the lateral cell membranes. 8 μm/grid square.* 6b. *15 minute incubation of rat duodenum showing intracellular calcium localizations in a supra-nuclear position and clearly separate from the lateral cell membranes. 5 μm/grid square.*

minutes revealed discrete localizations of calcium within the absorptive cells near the apical surface, as shown in Figure 4. These localizations were usually

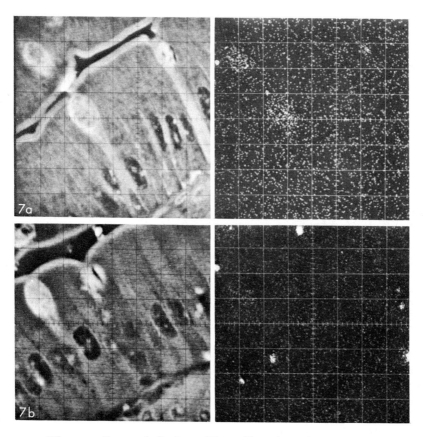

Figure 7a. *Calcium distribution in vitamin D-depleted rat. No calcium localizations occur in absorptive cells in control or incubated tissue. 8 μm/grid square. 7b. Calcium distribution in vitamin D-repleted rat. 15 minute incubation. Frequent calcium localizations are observed within absorptive cells. Although not shown here, increased amounts of calcium were observed within goblet cells. 8 μm/grid square.*

near the lateral cell borders at the level of the junctional complex. In traversing the cell, most of these localizations appear to remain associated with these lateral membranes, as illustrated in Figure 5a. After 15 minutes of incubation the calcium localizations in the rat were at the level of the cell nucleus and again appeared to be associated with lateral membranes or the intercellular space. In the chicken the localizations appeared to move more rapidly and after 15 minutes of incubation were frequently associated with the lateral membranes or intercellular spaces below the nuclei, or even in the lamina propria (Figure 5b).

Although the majority of calcium localizations within the absorptive epithelium were near lateral membranes, some localizations were observed within the cell cytoplasm clearly separate from these lateral membranes. Apical localizations observed after five minutes of incubation are shown in Figure 6a. More easily demonstrated are the localizations within the cytoplasm observed above the cell nucleus. This distribution is observed after 15 minutes of incubation and is shown in Figure 6b. Localizations were not observed within nuclei or in the cytoplasm below nuclei.

In vitamin D-deficient animals fed *ad libitum*, absorptive cells were devoid of calcium localizations, although calcium was still found in association with goblet cell mucus, as shown in Figure 7a. Incubation in a 2 mM calcium solution had no effect on this calcium distribution; in particular, calcium localizations within the absorptive cells were non-existent.

Vitamin D-replete animals fed *ad libitum* exhibited a calcium distribution different from that described for both D-deficient and normal animals. Calcium was still associated with goblet cell mucus, but in greatly increased amounts; in addition, localizations within absorptive cells were commonly observed. After incubation in a 2 mM calcium solution there was a dramatic increase in the number of these calcium localizations within absorptive cells (Figure 7b). The rate of translocation and the morphological distribution of these localizations were the same as that observed in normal tissue.

In order to determine if some calcium transport process not represented by localizations was ocurring, but was escaping detection due to the low sensitivity of scanning images, integrated point counts were taken with the electron beam on the apical cytoplasm. Cytoplasmic calcium was detected by this technique; however, there was no change in the cytoplasmic level of calcium in tissue from normal, vitamin D-deficient or D-replete animals, nor was there a difference in cytoplasmic calcium levels between control or incubated tissue.

DISCUSSION

There are several points which can be made from this microprobe study. One concerns the path of calcium translocation. One means by which transporting cells might avoid increasing their cytoplasmic activity of this potentially toxic ion, is to have the transported calcium move around rather than through the cells. Unfortunately, the resolution inherent in these studies does not permit a definitive answer, but the fact that the majority of the calcium localizations were in the vicinity of the lateral cell membranes is compatible with such a process.

Although many of the calcium localizations could conceivably be located between cells, some localizations were clearly within the cell cytoplasm (Warner 1972). The discrete nature of all of the calcium localizations indicates that this ion is in some way sequestered during transport. Within the cell, a mechanism which compartmentalizes the transported calcium is clearly advantageous to cells which must handle bulk quantities of this ion.

R. R. Warner and J. R. Coleman

A third and major point derived from this study concerns the nature of the compartmentalization process. The association between calcium localizations and both intra- and extracellular goblet cell mucus suggests a role for this mucus in the calcium transport process. This association is further supported by similarities in the form of calcium localizations both within goblet cells and in transit across absorptive cells, by the nature of the vitamin D response in goblet cells, and by similarities among mucus, calcium localizations, and the cellular distribution of CaBP.

In conclusion, this microprobe study has provided new insight into the cellular mechanism of calcium transport. The possibility of goblet cell involvement in any transport process has previously not been given serious consideration. The results presented here suggest that calcium is transported by a sequestering process that involves goblet cell mucus, and indicate that biochemical isolation and homogenization studies, and tracer flux studies, should be undertaken to test this possibility.

The complementary nature of conventional and microprobe studies has previously been mentioned. Although conventional techniques are admirably suited for transport characteristics of a tissue, they are ill-suited to determine the actions of individual cells The reverse is true of microprobe studies. Informatior at both levels is of value. For instance, conventional techniques, which average results over a population of cells, would not have detected an interaction between goblet and absorptive cells. The interplay between these techniques promises to have a significant effect on the study of epithelial transport mechanisms

ACKNOWLEDGEMENTS

This paper is based on work performed partially on NIH Grant No. AM-14272 and NIH Training Grant

No. 5-TIGM-1088, and partially under contract with the U.S. Atomic Energy Commission at the University of Rochester Atomic Energy Project and has been assigned Report No. UR-3490-311.

REFERENCES

Borle, A. B., 1967, *Clin. Orthoped. and Rel. Res.*, 52, 267-291.

Chance, B., 1963, *Energy-Linked Functions of Mitochondria*, Academic Press, New York.

Coleman, J. R., and Terepka, A. R., 1972a, *J. Histochem. Cytochem.*, 20, 401-413.

Coleman, J. R., and Terepka, A. R., 1972b, *J. Histochem. Cytochem.*, 20, 414-424.

Harrison, H. E., and Harrison, H. C., 1966, *Proc. Soc. Exp. Biol. Med.*, 121, 312-317.

Kachmar, J. F., and Boyer, P. D., 1953, *J. Biol. Chem.* 200, 669-682.

Kimmich, G. A., and Rasmussen, H., 1969, *J. Biol. Chem.* 244, 190-199.

Martin, D. C., and DeLuca, H. F., 1969, *Arch. Biochem. Biophys.*, 134, 139-148.

Taylor, A. N., and Wasserman, R. H., 1970, *J. Histochem. Cytochem.*, 18, 107-115.

Warner, R. R., and Coleman, J. R., *Microprobe Analysis of Calcium Transport in the Small Intestine.* Submitted for publication.

Warner, R. R., 1972, Ph. D. Thesis, University of Rochester, Rochester, New York.

R. R. Warner and J. R. Coleman

Wasserman, R. H., 1968, *Calc. Tiss. Res.*, <u>2</u>, 301-313.

Wasserman, R. H., and Taylor, A. N., 1966, *Science*, <u>152</u>, 791-793.

Wasserman, R. H. and Taylor, A. N., 1969, <u>In</u>, *Mineral Metabolism*, vol. III, (Comar, C. L., and Bronner, F. Eds.), Academic Press, New York, 321-403.

DISCUSSION

QUESTION: Is it possible that all of the calcium localizations you observed were actually near the cell borders? Could the thickness of the section have deceived you into thinking some of the calcium localizations were intracellular?

ANSWER: Most of these sections were 1 micrometer thick; some were 2 micrometers. It is possible that portions of two cells might be contained within the section, and a localization within the intercellular space between the cells might then appear in the sample current image to be inside one cell. We tested this possibility by taking 1 micrometer thick sections transverse to the long axis of these columnar cells. Such sections clearly show the cell borders. Although most of the calcium localizations were in this border region, some localizations were clearly seen within the cells.

QUESTION: What are those white spots in the sample current image that are associated with the calcium localizations?

ANSWER: The sample current images that you saw are formed primarily by atomic number differences. Since calcium is atomic number 20, whereas the matrix is somewhere around atomic number 6, the calcium localizations will stand out and in this (reverse) image appear as white spots.

Intestinal Calcium Transport

QUESTION: Are you quite sure your results are related to the transport of calcium? Goblet cells extrude, not reabsorb. Could incubation be creating an overload of calcium with resulting calcification not related to transport?

ANSWER: There was no overloading of calcium due to incubation. The concentration of calcium in which this tissue was incubated is 2 mM. This is a low, physiological level. The technique used to prepare the tissue is borrowed from that of the muscle physiologists; oxalate is used to precipitate calcium. Calcification is not involved; it is a matter of precipitating calcium that was in that vicinity.

QUESTION: It is diffcult to study calcium in tissue, especially when it is localized in a very small area. The same applies to silicon and titanium. How can we be sure that what you detect in the probe is not dust from the atomosphere?

ANSWER: We did not observe a random distribution of calcium, either in normal or incubated tissue. The distribution we observed was a very specific localization in very specific places. Simply having calcium fall non-randomly on a tissue section is not possible.

QUESTION: From a physiological point of view, your results are very interesting in the light of what we know of sodium transport. There is evidence that part of the transported sodium passes not through the cell but around the cell between the adjoining membranes. Your slides show this for calcium also.

QUESTION: You made a very optimistic statement in the beginning of your talk that there is a preparative technique that prevents calcium redistribution. From muscle cell physiology we know that about 70% of cellular calcium is very easily exchanged with other compartments. With classical techniques of dehydration and embedding I am pretty sure you have redistributed most of your calcium. You didn't give many specifics about the technique. Could you describe your sample preparation?

ANSWER: The point you raised is critical — does the technique really work. Extensive investigations have been conducted to test this point, and I refer you to the more detailed discussion (J. Histochem. Cytochem. 20, 401-423 (1972); Warner, 1972). The preparative technique is as follows: the tissue is removed and placed in buffered acrolein fixative, an extremely fast acting fixative. The fixative also contains 1% oxalate to immobilize calcium. If you don't use acrolein but use osmium, a slow-diffusing fixative, then calcium is redistributed; there is no calcium within the cells, only outside the cells. If you omit oxalate, calcium cannot be found anywhere. After one hour of fixation, the tissue is post-fixed in osmium containing 1% oxalate, and is dehydrated in oxalate-saturated alcohol and propylene oxide. The tissue is then embedded in Epon, cut dry, and mounted without contact with water.

QUESTION: How much calcium is lost to the solution?

ANSWER: Experiments have been done to determine the amount of calcium lost to the solutions. After prior incubation in ^{45}Ca, one can process the tissue and look at what comes out. This was not done with the intestine but was done on the chick chorioallantoic membrane. The preparative procedure retains within the tissue 80% of the calcium. The 20% that is lost is accounted for in the first step, the acrolein fixation step. After that step there is no further loss of calcium.

QUESTION: If these calcium concentrations are near the nucleus and are there in any large fraction of cells, would it not be worthwhile to go to the transmission electron microscope to see what they are?

Intestinal Calcium Transport

ANSWER: People have attempted to use the electron microscope to see calcium transport for a long time. They have all been unsuccessful.

QUESTION: But you know where to look and what it should look like from your probe studies. What is there? Is there some structure there that is definable?

ANSWER: When one looks at intestinal cells, there are a lot of things there that one does not understand. There are multivesicular bodies, dark bodies, a lot of structures within the cell that simply are not understood. One can't say whether they are transporting calcium or not.

QUESTION: What bothers me is the extraordinary concretion. It looks to the layman almost like it was crystallized. It is very suggestive of calcium oxalate. I doubt if you can be sure that this is not an artifact. If you look at incipient ossification and anything else I know, it is much more diffuse than these particles.

ANSWER: If you look in the transmission electron microscope in regions in thick sections where calcium localizations are shown to occur, you find no crystalline material — that's in a 1 μm thick section. If you try to cut thinner sections you get very poor morphology because they have to be cut dry. If you cut a thinner section on water you remove the calcium. Whatever the form of the oxalate precipitate, it is a very fine precipitate. It is not large enough to show up under the limited resolution of thick sections in the transmission microscope or the poor resolution you also get in dry light microscope sections.

ELECTRON PROBE ANALYSIS OF CALCIUM-RICH LIPID DROPLETS IN PROTOZOA

J. R. COLEMAN, J. R. NILSSON[1], AND R. R. WARNER[2]
*Department of Radiation Biology and Biophysics,
University of Rochester School of Medicine
and Dentistry, Rochester, New York*

INTRODUCTION

Recent investigations have suggested that intracellular lipid accumulations may play an important role in biological mineralization (Anderson and Coulter, 1969; Kashiwa and Atkinson, 1963; Nichols *et. al.*, 1971). Certain protozoa, *Tetrahymena pyriformis* and *Amoeba proteus*, are known to form intracellular lipid-containing granules which are also rich in inorganic materials (Allison and Ronkin, 1967; Elliot and Bak, 1964; Elliot *et al.*, 1966; Levy and Elliot, 1968; Levy *et al.*, 1969; Mast and Doyle, 1935; Nilsson and Forer, 1972). In *T. pyriformis* these are termed "refractive granules", and "refractive bodies" in *A. proteus*. Since both of these organisms are readily cultured in the laboratory, they offer the possibility of studying some of the characteristics of intracellular calcium accumulation under defined conditions. However, bulk analysis of the granules has not been easy to accomplish (Rosenberg, 1966; Rosenberg and Munk, 1969). Because the inclusions contain both lipid and water soluble materials, isolation procedures

---

[1]Present address: Institut fur Almen Zoologi, V Kobenhaven 'Universitet, Kobenhaven, Denmark.
[2]Present address: Department of Physiology, Yale University School of Medicine, New Haven. Conn.

employing organic solvents or aqueous media are likely to selectively remove one or another component. Fortunately, electron probe analysis can be employed to measure the calcium contents of such organelles (Coleman *et al.*, 1972; Coleman *et al.*, 1973a; Coleman *et al.*, 1973b). We undertook to elucidate the metallic composition of these granules *in situ*, and to determine whether their composition could be manipulated by changing the composition of their nutrient medium.

METHODS — RESULTS

Preparing the organisms for analysis posed several problems. The surface of *T. pyriformis* contains protective organelles termed mucocysts. When the organism is disturbed, these mucocysts rapidly extrude a mucous substance which has a high affinity for cations (Nilsson, 1970). It stains readily with alcian blue, a cationic dye, and complexes metals from the nutrient medium. If the cells were handled with unnecessary roughness, the mucocysts discharged, extruding mucus, which complexed calcium, magnesium, potassium and sodium from the medium, and caused the cells to be coated with a layer of metal-containing mucus. Furthermore, both *T. pyriformis* and *A. proteus* have diameters greater than the depth of penetration of the 22kV electron beam we wished to employ. Thus it was necessary to section the organisms or, alternatively, to keep them intact but flatten them without damaging their integrity. Conventional fixation, dehydration, embedding and sectioning methods could not be expected to preserve both the lipid and water soluble components of the organelles. Freezing methods were next considered. Since freeze substitution involves exposure to organic solvents, some lipid loss, and concommitant loss of metal complexes with them, would be likely. Freeze drying was also attempted with *T. pyriformis*. Cells were freeze dried in two ways: a thin film of nutrient medium and cells was spread on

a silicon disc and immediately plunged into a bath of
propane or freon cooled by liquid nitrogen; or cells
were centrifuged gently, some medium decanted and the
centrifuge tube immersed in a propane or freon bath
cooled by liquid nitrogen. Both types of preparation
were dried at either liquid nitrogen temperature or
acetone-dry ice temperature for periods ranging from
two days to more than a week. Cells dried as a film
on a silicon disc seemed morphologically well preserved but were too thick to permit the identification
of refractile granules in sample current images.
Cells freeze-dried as a suspension showed evidence of
severe morphological distortion, and were never embedded for sectioning. Air drying at ambient temperatures was attempted but the cells remained too thick
for granules to be resolved in sample current images.
Finally heat fixation was attempted; the cells were
spread in a thin film on a silicon disc, most of the
medium was quickly removed by tilting the disc and
touching the edge with the torn edge of a filter paper,
and the disc was passed through the flame of a propane
torch. The resulting preparation contained a band of
flattened cells in which refractile granules were visible in sample current images. On either side of this
band cells were either too thick for sample current
resolution of granules, or flattened to the point
where their morphology was obviously distorted
(Coleman, 1972; Coleman *et al.*, 1973a; Coleman *et al.*,
1973b).

As with any histochemical method, the results of
electron probe analysis depend on whether the normal
in vivo distribution of elements was preserved prior
to analysis. However, in testing the validity of such
a preparation one is faced with several problems.
First, only some of the cells in any preparation will
be suitable for analysis, thus routine bulk methods of
analysis cannot be employed to test the preparative
procedure. Second, there are not feasible alternative
microanalytical techniques with comparable chemical
and spatial resolution, so that one is placed in the

logically uncomfortable position of using electron
probe analyses to test and corroborate other electron
probe analyses. Third, one must preserve the normal,
in vivo distribution of elements when this is unknown,
and is, indeed, the object of the study.

Fortunately, there are some criteria that the
preparations must meet if they are intact and have not
suffered major redistributions of elements, and whether cells meet these criteria can be tested with electron probe analysis (Coleman *et al.*, 1972; Coleman
et al., 1973a and Coleman *et al.*, 1973b). First, if
the cells have remained intact, then there should be
a high internal potassium content, and low sodium content. Second, the profiles of sodium and potassium
should fall off sharply at the cell boundary. If
there is a gradual and overlapping decrease in the
profiles at the cell boundary, then there is reason to
suspect that the plasma membrane was disrupted during
preparation, permitting sodium to diffuse from the
high sodium medium into the low sodium cytoplasm and
permitting potassium to diffuse from the high potassium cytoplasm into the low potassium extracellular
space. Third, if any intracellular structures contain
high concentrations of diffusible elements, then there
should be no evidence of a local depletion or enrichment in the cytoplasm adjacent to the structure. In
the case of a local depletion surrounding the structure, one would suspect that the elements diffused
into the structure; while in the case of a local enrichment surrounding the structure, one would suspect
that the element diffused out of the structure into
the surrounding cytoplasm. It is hardly necessary to
point out that if the cells in a preparation meet
these criteria, it does not prove that the original
distribution of elements was preserved, it means only
that the characteristics of the preparation are consistent with the cells having preserved their original
distributions.

In the case of cells dried as described above,

Ca-Rich Lipid Droplets in Protozoa

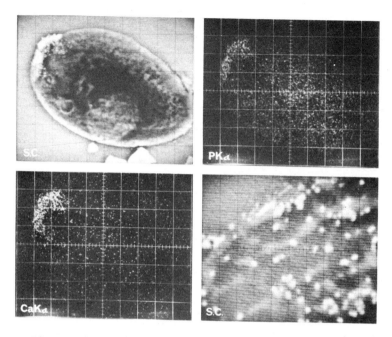

Figure 1. *Sample current and X-ray images of heat-fixed T. pyriformis. The sample current image, upper left (S.C.), shows a flattened cell at a magnification of 10 μm/screen division. A cluster of refractile granules appears as bright spots in the posterior region of the cell. The phosphorus X-ray image, upper right (PKα), and calcium X-ray image, lower left (CaKα) indicate that the region of cytoplasm containing granules is also rich in these two elements. A higher magnification, 2 μm/screen division, shows the appearance of granules in a sample current image, lower right, S.C.*

all three criteria were met. This can be seen in Figures 1 and 2. Here the sample current and calcium, phosphorus and potassium X-ray images are presented. It can be seen (Figure 2b) that K is restricted to the cells. Figures 3 and 4 show that there is no detectable depletion or enrichment in the cytoplasmic region surrounding the granules.

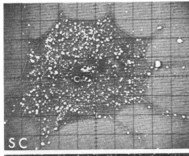
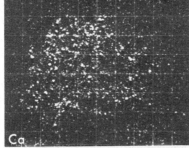
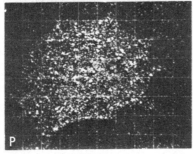

Figure 2a. *Sample current and X-ray images of heat fixed A. proteus. The sample current image (S.C.) shows the shape of the flattened cell, and the numerous refractive bodies in the cytoplasm. The calcium X-ray image (Ca) indicates that a major part of the cell calcium is located in the bodies, while the phosphorus X-ray image (P) shows that phosphorus is distributed throughout the cytoplasm, as well as within the refractive bodies.*
40 µm/screen division.

Tetrahymena pyriformis

For electron probe analysis, *T. pyriformis* were cultured in three different commonly employed media: proteose peptone (Plesner, *et al.*, 1964), medium M (Nilsson and Forer, 1972), and defined medium (Holz, *et al.*, 1959). In order to elucidate some of the effects of medium composition on the granules, two of the media were supplemented with two different concentrations of calcium and strontium. X-ray intensities were corrected according to the BICEP or BASIC correction procedures.

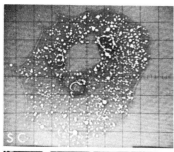

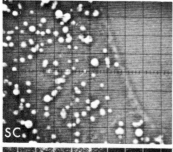

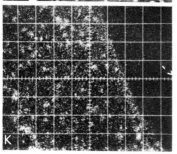

Figure 2b. *Sample current and X-ray images of heat fixed A. proteus. The uppermost image is a sample current image at a magnification of 40 µm/screen division. The shape of the flattened cell is apparent. Refractive bodies are distributed throughout the cytoplasm, and a large contractile vacuole occurs near the center of the cell. The middle image is the sample current image, S.C., of a portion of the same cell showing individual refractive bodies, and the edge of the cell at a magnification of 10 µm/screen division. The bottom image is the potassium X-ray image (K) from the same portion of the cell seen in the middle image. The edge of the potassium X-ray image coincides with the edge of the cell seen in the sample current image. This indicates that little or no leakage of cell potassium occurred during preparation.*

Analysis of individual refractile granules of *T. pyriformis in situ* showed that all granules contained substantial amounts of carbon. This was an expected result since the granules had been shown to be rich in lipid by histochemical staining methods (Levy and Elliott, 1968; Nilsson and Forer, 1972). Each granule also contained calcium, magnesium, phosphorus and potassium. Some contained small amounts of sodium (Coleman *et al.*, 1972; Coleman *et al.*, 1973b). The amount of calcium, magnesium, potassium and phosphorus contained by a granule was quite variable; but the

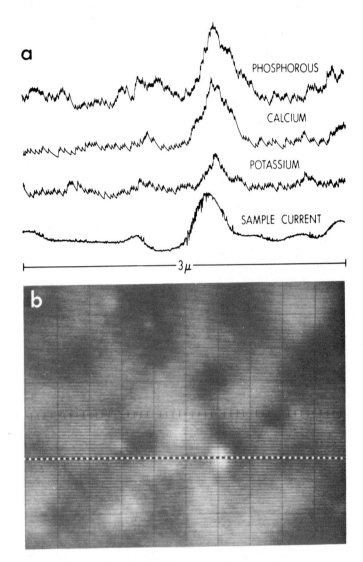

Figure 3. *Sample current image (b) and sample current and X-ray profiles of individual refractile granule in heat fixed T. pyriformis. The X-ray profiles were generated by scanning the beam along the path indicated by the dotted line. (The X-ray profiles are offset to the right of the sample current profile because of a long time constant in the X-ray*

atomic ratios of Ca/P, Mg/P, K/P and (Ca + Mg + K)/P
tended to be more uniform. This can be seen in Table
I where the mean ratios for granules in cells grown
in various media are presented. The relatively large
standard deviations associated with these mean ratios
give some indication of the range of values found in
each population. That the variation in ratios was not
due to the analytical technique was established by
analysis of a standard preparation of 0.264 μm
±(0.006 μm) diameter polystyrene latex particles
(Coleman *et al.*, 1972). The variation in carbon X-ray
intensity from these particles was found to correspond
closely to the measured size distribution, and the
calculated precision (standard deviation/mean) expressed as percent was between 1% and 2%. The accuracy of BICEP is about ±10% (Coleman *et al.*, 1972;
Warner and Coleman, 1973).

It was found that the Ca/P and Mg/P ratios were
not significantly different in any of the three commonly employed culture media. The K/P ratio in the
defined medium was significantly lower than in the
other two media. Supplementing proteose-peptone
medium with calcium or strontium produced a significant decrease in the mean Ca/P ratio. Only addition
of 3mM Sr caused a significant alteration in the mean
Ca/P ratio in medium M. In proteose-peptone, the
mean Mg/P ratio was elevated by supplementation with
0.3mM Ca but not by 3mM Ca. In contrast 3mM Sr decreased the mean Mg/P ratio, while in medium M 0.3mM
Sr decreased the Mg/P ratio. Supplementation of proteose-peptone media had no significant effect on the
mean K/P ratio, but 0.3mM Ca and 0.3mM Sr significantly increased the K/P ratio in medium M.

*rate meter). The X-ray profiles indicate that the
region around the granule has not been enriched nor
depleted of these elements. This is taken as evidence
that diffusion of these elements into the granules
from surrounding cytoplasm or from the granules into
the cytoplasm has not occurred.*

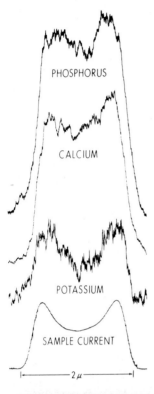

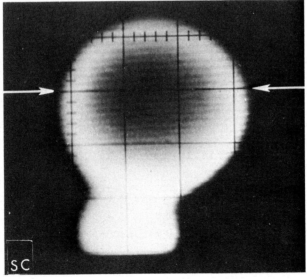

Figure 4. *Sample current image (S.C.) and sample current and X-ray profiles of individual refractive body in heat fixed A. proteus. The profiles were generated by scanning along the path indicated by the arrows. The shape of the X-ray profiles coincides with that of the sample current profile, indicating that there was no detectable loss or gain by the bodies during preparation.*

TABLE I

Mean Ratios of Atomic Percents of Metals to Phosphorus in Refractile Granules in
T. pyriformis Grown in Various Media with Different Divalent Cation Supplements.

Medium	Ca/P(±S.D.)	Mg/P(±S.D.)	K/P(±S.D.)	Sr/P(±S.D.)	Divalent Metals/P	N
1. "P.P"(10^{-4}M Ca; 10^{-3}M Mg)	0.44±0.06	0.33±0.12	0.12±0.02	–	0.73±0.10	40
2. P.P. + 0.3 mM Ca	0.29±0.06[a]	0.49±0.09[d]	0.12±0.04	–	0.78±0.10	4
3. P.P. + 3.0 mM Ca	0.23±0.08[a]	0.48±0.09	0.12±0.03	–	0.71±0.10	16
4. P.P. + 0.3 mM Sr	0.33±0.06[a]	0.29±0.06[d,e]	0.10±0.06	0.15±0.02	0.70±0.12	4
5. P.P. + 3.0 mM Sr	0.23±0.08[a]	0.27±0.06[d,e]	0.09±0.05	0.28±0.08	0.70±0.08	19
6. "M" (10^{-4}M Ca; 10^{-3}M Mg)	0.41±0.06	0.31±0.10	0.14±0.06[g]	–	0.73±0.10	40
7. M + 0.3 mM Ca	0.39±0.09	0.30±0.05	0.19±0.03[g]	–	0.70±0.10	5
8. M + 3.0 mM Ca	0.50±0.10	0.26±0.08[f]	0.08±0.06	–	0.71±0.10	4
9. M + 0.3 mM Sr	0.39±0.20[b,c]	0.26±0.02[f]	0.20±0.02[g]	0.08±0.02	0.69±0.01	5
10. M + 3.0 mM Sr	0.28±0.13[b,c]	0.27±0.04	0.19±0.06	0.29±0.06	0.75±0.10	8
11. "D"(7.5×10^{-4}M Ca; 2×10^{-3}M Mg)	0.43±0.06	0.31±0.02	0.10±0.04[h]	–	0.72±0.10	40

a. Significantly different from medium 1 at less than 0.05 level (t-test)
b. " " " " " " " " " 6
c. " " " " " " " " " 8
d. " " " " " " " " " 1
e. " " " " " " " " " 3
f. " " " " " " " " " 6
g. " " " " " " " " " 1
h. " " " " " " " " " 1 and 6 "

In media containing strontium all granules contained strontium, and the mean Sr/P ratio was higher
in cells grown in 3mM Sr than in cells grown in 0.3mM
Sr. It is not clear that Sr competed with only a single element for incorporation. For example, in pro-
teose-peptose medium 0.3mM Sr significantly decreased
the Ca/P ratio while the Mg/P ratio showed a decrease
that was not statistically significant. Inclusion of
3mM Sr significantly decreased both Mg/P and Ca/P.
This effect was not found in medium M. Here, 0.3mM
Sr significantly decreased the Mg/P ratio, and produced no significant decrease in the Ca/P ratio, while
3.0mM Sr had the opposite effect of producing a significant decrease in the Ca/P ratio, and none on the
Mg/P ratio. From these observations it is not possible to determine whether the effects of supplementing
the medium are due to the increase in concentration
of a specific ion, or just the consequences of an increase in the divalent ion concentration.

Amoeba proteus

In order to extend these observations we examined
another protozoan, *Amoeba proteus* which contained morphologically similar granules (termed "refractive
bodies"). Histochemical investigations had shown that
these bodies were phospholipid and contained inorganic
materials (Byrne, 1963; Heller and Kopac, 1956). Individual *A.proteus* were prepared in the way described
for *T. pyriformis* and tested according to the criteria
outlined above for the possibility of redistribution
(Coleman *et al.*, 1973a). X-ray and sample current
profiles can be seen in Figures 2 and 4.

The refractive bodies of a proteus have two different forms. The largest ones have a center that
appears hollow while the smaller ones appear solid
(Figure 3). We compiled the analyses according to
size and morphology, and these can be seen in Table
II. As in *T. pyriformis* each body contained a variable amount of calcium, magnesium, potassium,

phosphorus and carbon. Some granules contained traces of sodium and/or chlorine. Also, as in *T. pyriformis* the Ca/P, Mg/P, K/P, and (Ca + Mg)/P ratios tended to be uniform. With one exception each class of granules was relatively similar in Ca/P, Mg/P, K/P, and (Ca + Mg)/P ratios. The exception is the Ca/P ratio of the smallest refractive bodies which is significantly different from the Ca/P ratio in refractive bodies of other sizes.

Isolated Granules

Refractive granules were isolated by previous workers from *Tetrahymena pyriformis* grown in a defined medium (Rosenberg, 1966; Rosenberg and Munk, 1969). Chemical analysis of a population of isolated granules showed them to be mostly calcium, magnesium and phosphate. There were traces of sodium and potassium, and no organic matter. These authors concluded that the granules were an intracellular accumulation of calcium and magnesium phosphates. They found that the composition of the granules could be altered by changing the divalent ion content of the medium, and some of their results are presented in Table III. It can be seen that with constant Mg concentration, a one hundred fold decrease in Ca produced only a 22% difference in the Ca/P ratio, and no significant change in the Mg/P ratio. At constant Ca concentration, decreasing the Mg concentration by about one third increases the Ca/P ratio by 30% and decreases the Mg/P ratio by 37%. This result would suggest that the Mg/P ratio is relatively independent of the calcium concentration in the medium, while both Mg/P and Ca/P ratios of the granules respond to changes in the medium concentration of magnesium. These authors also found that neither calcium nor magnesium alone was sufficient to permit granule formation, and that strontium could substitute for either element. The amount of strontium incorporated into granules was not reported, however.

TABLE II

Mean Ratios of Atomic Percents of Metals to Phosphorus in Refractive Bodies in *A. proteus*

	Ca/P(±S.D.)	Mg/P(±S.D.)	K/P(±S.D.)	(Ca + Mg)/P	N
Large					
1. Edge	0.33±0.07	0.12±0.05	0.22±0.06	0.45±0.05	26
2. Center	0.38±0.06	0.10±0.06	0.18±0.04	0.47±0.05	21
3. Intermediate	0.35±0.01	0.10±0.03	0.23±0.05	0.45±0.02	28
4. Small	0.25±0.05*	0.13±0.05	0.25±0.06	0.38±0.04	25

*Significantly different from 1, 2, and 3 at less than 0.01 level (t-test).

TABLE III

Molar Ratios of Divalent Metals to Phosphorus in Granules Isolated from *T. pyriformis* (from Rosenberg and Munk, 1969).

Medium	Ca/P	Mg/P
10^{-3} M Ca; 1.6×10^{-3} M Mg	0.50	0.51
10^{-5} M Ca; 1.6×10^{-3} M Mg	0.39	0.55
10^{-3} M Ca; 0.6×10^{-3} M Mg	0.65	0.32

Ca-Rich Lipid Droplets in Protozoa

Electron probe analysis of individual granules confirms and extends these findings. For example, the isolated granules had lost potassium and some phosphorus, while the analysis *in situ* shows potassium is a consistent component of granules and more phosphorus is present than previously reported. The isolated granules were free of organic matter while those *in situ* were rich in carbon — up to 70% atomic percent. Furthermore the variation among individual granules in a population was not evident by bulk analyses of populations.

The fact that the composition of the granules can be altered by changing the composition of the medium confirms the work of Rosenberg and Munk (1969). It is also of interest that the effects of changing the metal composition of the medium are influenced by other properties of the medium. These findings are consistent with the suggestion that the granules may act as an intracellular metal buffer. These organisms live in fresh water, and it may be an advantage to store these useful metals in the form of a complex with lipid, in effect storing them within the cytoplasm but in a form in which they do not affect cytoplasmic properties. The similar granules in *A. proteus* may also serve as intracellular metal stores. Since this organism also lives in a fresh water environment and derives its nutriment from ingesting other organisms, it may live in a "feast or famine" economy. That is, after digesting an organism the concentration of metals could rise to a level that might interfere with normal cell function. Since essential metals would not be available when the organism was fasting, deleterious effects could occur from the lack of metals. A mechanism that could serve as a metal buffer would permit relatively large amounts of these elements to be stored intracellularly without affecting normal cell operation.

Finally it is worth noting that these lipid and phospholipid inclusions do not represent mineralized

inclusions such as reported by Pautard (1970), nor are
they just calcium and magnesium phosphates as reported
on the basis of bulk analysis of isolated granules.
Instead they are more similar to the lipid containing
"matrix vesicles" reported to occur in association
with bone and tooth formation (Anderson and Coulter,
1969; Kashiwa and Atkinson, 1963; Nichols *et al.*, ·
1971; Termine, 1972). If this is true then it may be
that this primitive cellular mechanism for handling
metals may have evolved into the mechanisms involved
in the formation of mineralized tissues by more complex organisms.

ACKNOWLEDGEMENTS

This report is based in part on work performed
under contract with the AEC at the University of
Rochester Atomic Energy Project and assigned Report
Number UR-3490-346 and in part on work supported by
USPHS Research Grant AM14272.

REFERENCES

Allison, B. M., and Ronkin, R. R., 1967, *J. Protozool.*
14, 313.

Anderson, H. C., and Coulter, P. R., 1969, *J. Cell
Biol.*, 41, 59.

Byrne, J. M., 1963, *Quart. J. Micr. Sci.*, 104, 445.

Coleman, J. R., Nilsson, J. R., Warner, R. R., and
Batt, P., 1972, *Exptl. Cell Res.*, 74, 207.

Coleman, J. R., Nilsson, J. R., Warner, R. R., and
Batt, P., 1973a, *Exptl. Cell Res.*, 76, 31.

Coleman, J. R., Nilsson, J. R., Warner, R. R., and
Batt, P., 1973b, *Exptl. Cell Res.*, 80, 1.

Elliott, A. M., and Bak, I. J., 1964, *J. Cell Biol.*, 20, 113.

Elliott, A. M., Travis, D. M., and Work, J. A., 1966, *J. Exptl. Zool.*, 161, 177.

Heller, I. M., and Kopac, M. J., 1956, *Exptl. Cell Res.*, 11, 206.

Holz, G. G. Jr., Erwin, J. A., and Davis, R. J., 1959, *J. Protozool.*, 6, 149.

Kashiwa, H. K., and Atkinson, W. B., 1963, *J. Histochem. Cytochem.*, 11, 258.

Levy, M. R., and Elliott, A. M., 1968, *J. Protozool.*, 15, 208.

Levy, M. R., Gollon, C. E., and Elliott, A. M., 1969, *Exptl. Cell Res.*, 55, 295.

Mast, S. O., and Doyle, W. L., 1935, *Arch. Protistenk.* 86, 279.

Nichols, G. Jr., Hirschman, P., and Rogers, P., 1971, In, *Cellular Mechanisms for Calcium Transfer and Homeostasis*, G. Nichols and R. Wasserman, Eds., Academic Press, New York, p. 211.

Nilsson, J. R., 1970, *Compt. Rendus. Trav. Lab. Carlsberg*, 38, 107.

Nilsson, J. R., and Forer, A. J., 1972, *J. Protozool.*, suppl 19.

Pautard, F. G. E., 1970, In, *Biological Calcification, Cellular and Molecular Aspects*, H. Schraer, Ed., Appleton-Century-Crofts, New York, p. 105.

Plesner, P., Rasmussen, L., and Zeuthen, E., 1964, In, *Synchrony in Cell Division and Growth*, E. Zeuthen, Ed., Interscience, New York, p. 543.

Rosenberg, H., 1966, *Exptl. Cell Res.*, <u>41</u>, 397.

Rosenberg, H., and Munk, N., 1969, *Biochim. Biophys. Acta.*, <u>184</u>, 191.

Termine, J., 1972, *Clin. Orthop. and Rel. Res.*, <u>85</u>, 207.

Warner, R. R., and Coleman, J. R., 1973, *Micron*, <u>4</u>, 61.

DISCUSSION

GULLASCH: I have a question concerning the preparation procedure. You mentioned that manipulations such as centrifugation changed the intracellular electrolyte content of *Tetrahymena*. Why do such manipulations as flame fixation and irradiation with high beam currents not cause redistributions?

COLEMAN: First, the change I mentioned is extracellular and due to the release of mucus from mucocysts. We expect some mass loss occurs. With regard to redistributions, we do not know what is happening in the rest of the cytoplasm outside the granule. We have no way of applying your criteria to material mounted on an opaque substrate. Our redistribution criterion for granules is based on the following line of reasoning. It is generally assumed that the concentration of potassium and calcium throughout the watery part of the cell, excluding organelles, is uniform. The refractive granules are normally suspended in this watery phase, and have a different concentration of both potassium and calcium than the surrounding cytoplasm. During drying the granules can: a) lose these elements to the surrounding cytoplasm;

b) gain them from the surrounding cytoplasm; or c)
suffer no detectable redistribution of these elements.
If a) were to occur then the cytoplasm adjacent to
the granules would be enriched in these elements, and
in the X-ray images a "halo" of increased concentration would surround the granules. This would also
cause the X-ray profile of a granule to be substan-
tially broader than the sample current profile. If
b) were to occur, then in the adjacent cytoplasm,
there would be a local depletion of these elements,
and the surface of the granules would be enriched.
This would appear as a "dark halo" around the granules
in X-ray images. The X-ray profile would also show a
"valley" around the granule and the sample current
profile would not be congruent with the X-ray profile.
If c) were to occur, then the X-ray images and profiles would be similar and congruent to the sample
current images and profiles.

KAUFMANN: I feel that the procedure of flame fixation
may introduce such complete redistribution of ions
that you cannot expect such particular kinds of small
displacement of ions. I think you will find complete
redistribution of water soluble compounds throughout
the whole cell. What you will have left are those
ions strongly bound to organic material. Furthermore
with the high beam current you use, you certainly induce a strong loss of mass during the period of analysis. We hardly know whether the figure you gave
from the quantitative analysis has something to do
with the living cell.

INGRAM: Several of us have stated during the past
few days that shock freezing is the most promising
method of preparing tissues and cells for analysis;
yet you have chosen not to pursue this technique.
Perhaps you would care to comment on your reasons for
doing so.

COLEMAN: I don't think there is time for a complete
comparison and critique. I think, however, we are

all faced with the problem of finding independent methods to assay for redistribution in preparative methods. With our flattened cells we can't use bulk studies. We must look for the best tests feasible for individual cells. At the time we began, the best test was the congruence of sample current and X-ray profile. If you have another I would be more than happy to hear it.

WARNER: If I may respond to Dr. Kaufmann, in the light microscope one can see these granules. They are at the level of light microscope visualization. They occupy only a portion of the cell and when these are dried down the background in the portion where the granules are is the same as the background at the opposite end, some 80 or more micrometers away. Consequently, there is no visible gradient going toward or away from these granules and it hardly seems likely that there could be a redistribution over that length of the cell.

ECHLIN: How long is *Tetrahymena*, about 100 μm?

COLEMAN: Yes.

ECHLIN: I would have thought you would have adopted the technique of Bachmann of spray freezing, where he put microdroplets into a cryogen. You certainly won't do worse and you might even do better.

COLEMAN: We are aiming at frozen sectioning the cells. At the time we began, flattening was necessary in order to see the droplets.

FÜLGRAFF: I think as long as you are interested in structure-bound inorganic material there is no reason you should not prepare cells the way you have.

MUIR: What happens if you just air dry the *Tetrahymena?*

Ca-Rich Lipid Droplets in Protozoa

COLEMAN: They are too thick to resolve individual refractile granules.

MUIR: What about the *Amoeba?*

COLEMAN: They break open, and the potassium leaks from the cell.

RUSS: The test you describe would only reveal redistribution over relatively small dimensions; if redistribution were occurring over gross dimensions this test would not reveal it.

COLEMAN: In order to distinguish gross from fine redistribution one can visualize two situations. In the case of fine scale redistribution, consider what would happen between two adjacent granules. Each granule has a much higher concentration (of calcium, magnesium, potassium and phosphorus) than the surrounding cytoplasm, and if redistribution occurred, one would expect a loss from the granule into the cytoplasm. In the region between two granules, material from both granules would accumulate, creating a higher concentration in this region than in those regions where the material was free to diffuse a great distance, (where there was no adjacent granule). If any substantial loss of this sort occurred, it should be evident in X-ray profiles as asymmetry. Since the granules accumulate at one region of the cell, one would expect long range redistributions to appear as concentration gradients downhill from the granule region to the opposite end of the cell. None were found.

RUSS: Yes, but you are assuming a diffusion type situation and that all the calcium in the granule is of exactly the same kind bound in exactly the same way, and will react in the same way. I see no reason that, for example, 70% of the calcium might not stay and 30% leave, and the part that is left could go quite far away.

COLEMAN: The fact that the sum of calcium, magnesium and potassium was always proportional to the phosphate suggests that they are bound in the same way.

WARNER: I would like to comment on the mass loss. In the case of BICEP, mass loss does not greatly affect the calculations because one assumes that everything not actually analyzed is "carbon", and that oxygen is present as phosphate, thus proportional to phosphorus.

RUSS: This is also due to the fact that you are dealing with elemental ratios.

GULLASCH: What beam current do you use?

COLEMAN: About 3 nanoAmps. In reference to the mass loss mentioned by Hall, the fast loss may well occur, and if so, we won't be able to detect it. If there is also a slow loss it must be very slow. We have repeated analyses on the same granules after exposing them to a scanning beam for an hour and have found no difference.

GULLASCH: Then perhaps the loss was in the first few seconds.

COLEMAN: Yes, it may be so. If so it is a loss of "mass" which is probably carbon, and not of the calcium, magnesium, potassium and phosphorus that we are most interested in.

HALL: Yes, a mass loss makes no difference to the measurement of elemental ratios.

KAUFMANN: However, I feel it might be useful to have some idea about the concentration of these elements in the organic material as well as elemental ratios, and of course if you wished to measure this you would run into trouble. It seems to me that with this preparative procedure what you will measure is the ratio of firmly bound calcium, firmly bound potassium, and firmly bound magnesium to firmly bound phosphorus. Potassium is potentially more troublesome because it is more mobile.

COLEMAN: Because of this mobility it is remarkable that you see such a strong potassium signal from the granules. It shows that there has not been a major movement down a very steep gradient from the granule to the surrounding medium.

KAUFMANN: I don't think that argument applies. If there is time for the potassium to move long distances during fixation, then you will find a practically homogeneous distribution of potassium throughout the cytoplasm and will not detect the criteria you describe.

COLEMAN: However, we do not see a homogeneous distribution of potassium. The situation you describe is just the opposite of what we see. We see local concentrations of potassium. We do not see an equilibrium situation high in entropy, we see a lack of entropy — a concentrated potassium distribution.

KAUFMANN: I am saying that what remains in the granule is a fraction of firmly bound potassium. The bulk of physiological investigations has shown cell potassium to be water soluble.

COLEMAN: The three cations equalled the amount of phosphorus so it is reasonable to conclude that potassium was bound, but a potassium phosphate is not a tight bond — it is extremely soluble, and probably exists as a counter ion for the phosphate.

SPURR: Are these particles birefringent before preparation?

COLEMAN: No, and in the electron microscope they do not exhibit any lamellar structure.

INTRACELLULAR Na$^+$ - K$^+$ CONCENTRATION OF FROG SKIN AT DIFFERENT STATES OF Na-TRANSPORT

A. DORGE, K. GEHRING, W. NAGEL AND K. THURAU
Department of Physiology,
University of Munich, Germany

INTRODUCTION

The epithelium of the frog skin, like other epithelial structures such as exocrine glands, intestine, or the kidney, has the property of actively transporting Na. Because the frog skin is readily available and is an easily handled preparation, it has often been used to study the mechanism of transcellular Na transport. In order to analyze the individual steps of this transcellular transport, it is necessary to localize the Na transport compartment, and to determine its electrolyte concentrations.

Until recently, it has only been possible to measure the intracellular electrolyte concentrations indirectly by chemical analysis. Using this method the total volume and the total Na content of the tissue is determined, and corrected for the extracellular volume and electrolyte content with the aid of an extracellular marker. This method has the disadvantage that extracellular markers do not provide an adequate estimate of the extracellular space (Cereijido *et al.*, 1968). Furthermore, only the average intracellular concentration of all cells, which individually may have widely differing values, can be obtained. Such a situation exists in the frog skin, which has a relatively complex structure consisting of several cell layers, which may differ in function.

The present paper deals with experiments in which
the electrolyte concentrations within single cells of
the frog skin epithelium were determined. Variation
of the experimental conditions should enable one to
obtain information about the epithelial cell layer
which is directly involved in transcellular Na transport.

EXPERIMENTAL PROCEDURE

The experiments were carried out using a scanning
electron microscope (Cambridge) with an energy dispersive x-ray detector (Nuclear Diodes). In order to
obtain maximum sensitivity of the detector system, the
x-ray detector was positioned less than 1 cm away from
the specimen.

To minimize dislocation of the electrolytes during
the preparation, the frog skin was shock-frozen in
cooled isopentane at $-140°C$ after the functional state
of the skin was determined by measuring the short cir-
cuit current in an USSING type chamber. The different
steps in the preparation of freeze-dried sections are
summarized in Figure 1. A small frozen piece of frog
skin was mounted on a cooled brass holder and inserted
into a Reichert cryomicrotome. The temperature in the
cryochamber was maintained at $-70°C$. The skin was
trimmed into a pyramid with the aid of a U-shaped notch
ground into the steel knife. After trimming, the skin
was cut into sections 2-3 μ thick. In order to prevent
sections curling, the sections were forced into a 5 μ
slit between the back of the knife and a small glass
plate, which could be maneuvered from outside the cryochamber with a micromanipulator. A ribbon of sections
was picked up from the back of the knife with a 1000 A
thick collodium foil mounted over a carbon tube. The
carbon tube, with the collodium film and sections attached, was inserted upside down into a cylindrical
hole in 2 brass plates between which a second collodium
film had previously been positioned. Thus, immediately

338

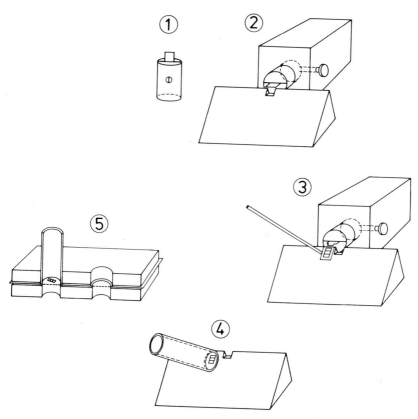

Figure 1. *Equipment used for preparation of frog skin sections. (1) frozen frog skin on plate, (2) and (3) trimming and cutting section, (4) removing section, (5) sandwiching section.*

after cutting, the sections were sandwiched between 2 collodium films. All these manipulations were performed in the cryochamber. The 2 brass plates also served as a holder for the carbon tube during the freeze-drying process.

These procedures may result in the formation of ice crystals which distort the morphology of the tissue and perhaps the location of the electrolytes. In order to determine the magnitude of this artefact,

339

frozen albumin solution was sectioned so that part of
the outer region, which had come into direct contact
with the isopentane, was included in each section.
Figure 2 shows an electron transmission image of such

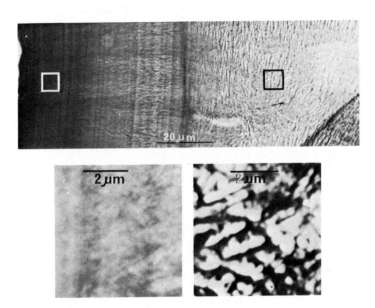

Figure 2. *Scanning transmission image of a
freeze-dried section of albumin solution. Enlarge-
ments of outer area (left) and inner area (right).*

a section. It can be seen that the albumin appears to
be distributed homogeneously throughout a surface layer approximately 30 μ thick. In deeper regions ice
crystal formation seems to have occurred to such an
extent that only a network of albumin remains. A
similar result was obtained by Christensen (1971).

Because the epithelium of the frog skin is relatively thin, ice crystal formation should not disturb
the morphological structures to a great extent. Figure
3 shows a scanning transmission image of a freeze-
dried section of the epithelium of the frog skin. The
epithelium, which consists of several distinct layers,

corneum, strat. granulosum, spinosum (2 layers) and germinativum, is approximately 30 μ thick, which corresponds to approximately 20% of the whole thickness of the frog skin. The corium which corresponds to approximately 80% of the thickness of the frog skin is

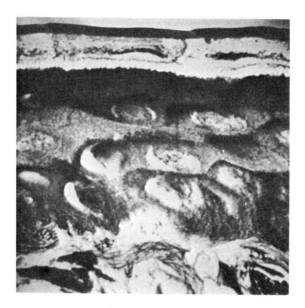

Figure 3. *Scanning transmission image of a freeze-dried section of the epithelium of the frog skin.*

only partly seen in this picture. In each layer areas of 1 - 4 μ^2 were scanned for 400 sec and the emitted x-rays were analysed in the energy range between 0.8 to 4 KeV, which contains the Kα-lines of the elements from Na to Ca. The Bremsstrahlung in this energy range is a measure for the mass content of the excited volume.

RESULTS AND DISCUSSION

In a first set of experiments, the intracellular Na and K concentrations in the different layers of the frog skin were measured under control conditions (incubation with frog Ringer solution on both sides of the skin) and after inhibition of the transcellular Na transport by ouabain (10^{-4} M in the bathing solution of the corium side). The mean intracellular concentrations for Na and K of all cells of the epithelium were also estimated by chemical analysis using the extracellular marker inulin to correct for the extracellular space.

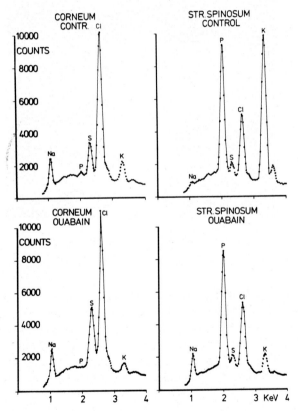

Figure 4. *Energy dispersive x-ray spectra obtained from the corneum and the stratum spinosum before and after ouabain. Counting time 400 sec. Beam current 5 × 10^{-10} A at 15 KV acceleration voltage.*

Na$^+$ — K$^+$ In Frog Skin

Figure 4 shows 2 pairs of x-ray spectra from measurements of the corneum and the strat. spinosum before and after exposure to ouabain. The Bremsstrahlung which constitutes the background of the spectra is practically the same, indicating nearly the same mass content in all cases. The Na peak in the spectrum obtained from the strat. spinosum after ouabain is considerably higher than that found under control conditions. On the other side the K-peak is smaller after ouabain. The peaks of the other elements P, S and Cl are not altered by ouabain. This effect of ouabain on Na and K peaks is typical for all epithelial layers except for the outermost layer of the epithelium, the corneum. In contrast to the spectrum obtained from the strat. spinosum, under control conditions; the spectrum of the corneum shows the typical extracellular pattern: a high Na and a small K peak which are not influenced by ouabain. The corneum, when compared to the cells of the strat. spinosum contains practically no P.

Further indications that the strat. corneum is in equilibrium with the extracellular space, comes from the observation that in nearly all experiments the Na, K and Cl concentrations were of the same order of magnitude as those found in the epithelial bathing solution. In addition, when the epithelial surface of the skin was rinsed for a short time (2 sec) with isotonic sucrose, the electrolyte content of the strat. corneum was reduced almost to zero. In some experiments (less than 10%), however, the concentrations of electrolytes in the strat. corneum were between those of the Ringer solution and the epithelial cells. The explanation for this phenomenon may be that these skins were in a particular phase of the cyclic moulting process which occurs every 30 hours.

For quantitation of the elements Na, and K, standards were prepared from albumin solutions cryosectioned and analyzed in the same manner as the frog skin. The albumin solution had a concentration of 22% which corresponds to the mean dry weight of the frog skin epithelium. Figure 5 shows calibration curves for Na and K. In the diagrams the ratio of characteristic

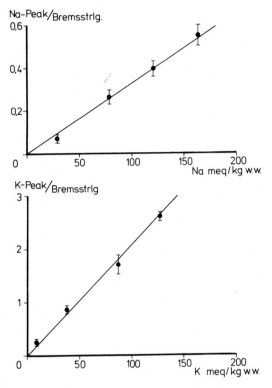

Figure 5. *Calibration curves for Na and K obtained from freeze-dried sections of a 22% albumin solution.*

x-rays to Bremsstrahlung is plotted against the Na and K concentration as determined by flame photometry. In both cases linear calibration curves were obtained.

The use of the ratio of the characteristic x-rays to Bremsstrahlung has the advantage of eliminating errors which originate from variations in the thickness of the sections, from instability of the beam current, and from alterations of the distance between

specimen and x-ray detector. A systematic error may
result from varying fractions of dry matter in biological structures as illustrated in Figure 6, which
shows 2 spectra from the same cell, one from the nucleus and the other from the cytoplasm. The peaks of

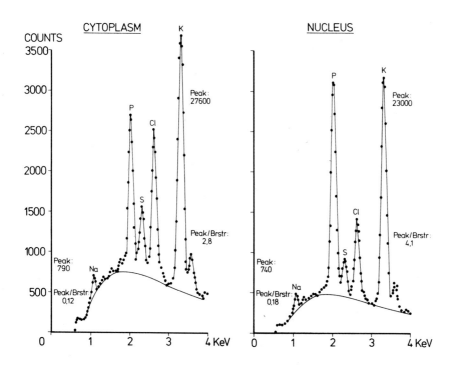

Figure 6. *Energy dispersive x-ray spectra obtained from the cytoplasm and nucleus of the same cell in
the stratum spinosum.*

Na and K are approximately equal. Because of the different fractions of dry matter, as indicated by the
different Bremsstrahlung, the ratio of peak to Bremsstrahlung would lead to the conclusion that the Na and
K concentrations in the nucleus are 1.5 times higher
than in the cytoplasm. However, if the difference in
dry matter is considered, the Na and K concentrations

in both intracellular structures are the same.

TABLE I

Comparison of Microprobe and Chemical Analyses
in meq/kg Wet Weight Before and After Ouabain.

	$[Na]_{ICF}$		$[K]_{ICF}$	
	Control	Ouabain	Control	Ouabain
Microprobe	35	120	135	43
Chem. Analysis				
a) whole skin	108	150	81	45
b) isol. epithelium	25	–	133	–

Values for isolated epithelium obtained from
Aceves and Erlij (1971).

Table I summarizes intracellular Na and K concentrations as determined by microprobe and chemical analysis before and after ouabain. The first row contains
the values of the microprobe analysis and the other
rows the data from chemical analysis of the whole skin
and of the isolated epithelium, which was separated
from the corium by the use of collagenase. Using the
microprobe, the Na content was calculated to be 35 and
120 meq/kg wet weight before and after ouabain respectively. After ouabain the K concentration decreased
from 135 to 43 meq/kg wet weight. The control values
measured with the microprobe are in good agreement
with the data which have recently been reported by
Aceves and Erlij (1971) for the isolated epithelium.
On the other hand the discrepancy between the microprobe data and those obtained by chemical analysis of
the whole skin is pronounced. We suggest that errors
in the estimation of the extra- and intracellular volume with chemical methods are the major source of this
discrepancy.

Further experiments were performed to obtain some
information about the localization of the Na transport

compartment. For this purpose the transcellular Na transport was reduced by the action of amiloride (10^{-4} M in the epithelial bathing solution) and by lowering the Na concentration in the epithelial bathing solution to almost zero. It is generally accepted that amiloride reduces the entrance of Na from the epithelial bathing medium into the frog skin and thus diminishes the unidirectional Na flux from the epithelial to the corium side. Should Na transport through the frog skin include a cellular transport compartment, amiloride would be expected to reduce the Na concentration in this compartment. A similar effect should be expected when the Na concentration in the bathing solution of the epithelial side is reduced.

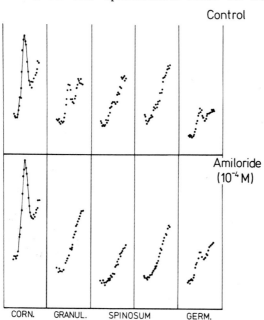

Control

Amiloride
(10^{-4} M)

CORN.   GRANUL.   SPINOSUM   GERM.

Figure 7. *Na. peaks obtained in the different layers of epithelium of the frog skin before and after amiloride.*

Figure 7 shows x-ray spectra of the different epithelial cell layers in the energy range of the Na peak under control conditions and after the action of amiloride. The Na peak in the corneum is not influenced by amiloride. In the control state, Na concentrations of 20 − 30 meq/kg wet weight were found in the strat. granulosum and germinativum. Na could not be found in the two layers of strat. spinosum. After the action of amiloride the Na content in the strat. granulosum decreased to almost zero, whereas in the strat. germinativum it remained unchanged.

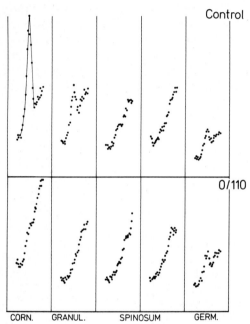

Figure 8. *Na peaks obtained in the different layers of epithelium in control and after incubating the epithelial side with Na-free solution and the corial side with Ringer's solution (110 meq/ l Na).*

In Figure 8 the effect of Na-free choline Ringer solution at the epithelial side and normal Ringer at the corium side upon the Na peaks are shown. The control spectra are similar to those in Figure 7. Na free Ringer on the epithelial side produced the same changes in the epithelial cells as amiloride. The Na concentration in the strat. granulosum is decreased whereas the concentration of the cells of strat.

germinativum is unaffected. The Na concentration in the corneum attains that of the epithelial bathing solution, which is nearly zero.

The identical decrease in Na concentration in the cells of strat. granulosum during both experimental procedures suggests that the Na in these cells is part of the Na transport pool. These data demonstrate that the use of the microprobe analysis is of great value in localizing changes in electrolyte concentrations of single cells which occur in the course of changes in electrolyte transport.

REFERENCES

Aceves, J., D. Erlij, 1971, *J. Physiol.*, 212, 195-210.

Cereijido, M., Reisin, I., Rotunno, C. A., 1968, *J. Physiol.*, 196, 237-253.

Christensen, A. K., 1971, *J. Cell. Biol.*, 51, 772-804.

ELECTRON PROBE MICROANALYSIS OF PICOLITER LIQUID SAMPLES

C. LECHENE

*Biotechnology Resource in Electron
Probe Microanalysis, Department of Physiology and
Laboratory of Human Reproduction and Reproductive
Biology, Harvard Medical School, Boston, Mass.*

INTRODUCTION

In 1923, G. C. de Hevesy with D. Coster discovered the element Hafnium, while studying the fluorescent radiation excited by x-rays; later he devised a method to calculate the relative abundance of the different chemical elements based on the analysis of charateristic x-ray lines. But, it was not until Castaing in his D. Sc. thesis lead by A. Guinier laid the practical and theoretical foundation of electron probe microanalysis, that x-ray microanalysis spread (Castaing, 1951). Since this time, electron probe microanalysis has come to be thought of as a powerful tool for biologists (Tousimis, 1962; Galle, 1965). Indeed, electron probe microanalysis could be applied in a wide range of biological research (Cosslett, 1966; Hall *et al.*, 1966). The first biological applications were on bulk, hard sample, such as bone, teeth, and foreign body inclusions; the biological samples are prepared and analyzed in the same manner developed for the original use of electron probe microanalysis in studying metallurgical or geological samples.

Great promise lies in the analysis of soft biological material, isolated cells, and thin or ultra thin sections, in order to map and quantitate the

351

elemental content of cells, intracellular organelles, and intercellular spaces. The difficulties being worked on are in the preparation of these biological samples for qualitative or quantitative analysis: we must avoid the redistribution of the element content during the preparation; we must study the effects of the interaction between the electron beam and the samples; we must know the volumes from which characteristic x-rays are emitted in order to get elemental concentration.

We would like to mention that electron probe microanalysis could be applied to study and quantitate the histochemical reactions which are used in light microscopy. Such reactions could also be studied in electron microscopic preparations where the reaction products could be recognized by their characteristic x-ray production; thus it will no longer be necessary to rely on an electron dense quality. A brief example will illustrate such an application in histochemistry. Gersh and Stieglitz (1934) studied kidney function by using sodium ferrocyanide precipitated by ferric chloride as the insoluble complex, prussian blue. They ascertained, by looking at the blue precipitate in histological kidney preparation, that sodium ferrocyanide remained extracellular. Using the same preparation, but complexing the sodium ferrocyanide with zinc chloride we can see (Figure 1) that iron and zinc are readily detected inside the epithelial cells of kidney proximal tubules. This indicates that part of the sodium ferrocyanide does penetrate the epithelial cells. We will not discuss here the significance of such an observation.

We would like now to describe another general method of wide possible application of electron probe microanalysis: i.e. to perform quantitative analysis of the elemental content of picoliter volumes of liquid. This method provides the biologist with a unique tool: it enables him to quantitate in the same minute picoliter sample each of the chemical elements

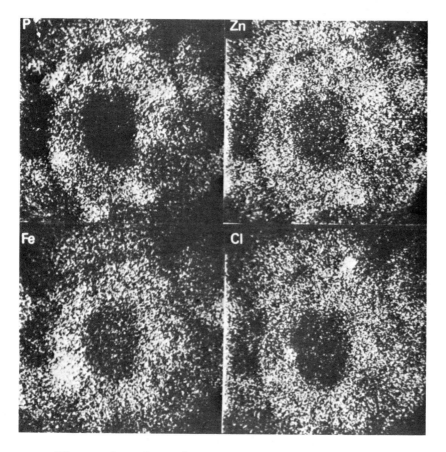

Figure 1. *Scanning x-ray of rat proximal tubule (see text). 30KV-80nA. Scanned area: 50 × 50 μ. Each picture integrated on 50,000 counts - (P: 650 sec, Zn: 1000 sec, Cl: 1300 sec, Fe: 3000 sec).*

whose atomic number is greater than five that are present in quantities even lower than 10^{-14}M. The elements which can be analyzed are from at least the lower atomic number of 6 (Carbon) up to the end of the periodic table. As the samples are not destroyed during the analysis we can analyze on the same sample all the elements in which we are interested. Characteristic x-ray lines are well defined and well parted by wavelength dispersive crystals. Thus, there are no interferences from one element to another.

Preliminary reports on the use of the electron
probe to analyze nanoliter liquid volumes have been
made by M. J. Ingram and Hogben in 1967, and by Morel
and Roynel in 1969; but the techniques described could
not be worked out routinely in the laboratory.

The principles of the method that we outlined
(Lechene, 1970) are the following: we dry a known
volume of liquid sample; we perform the x-ray analysis of the salt crystals with the electron probe; and
we compare the x-ray signals of the element in which
we are interested to the x-ray signal emitted from the
salt crystals obtained in the same condition with
standard solutions of known composition.

The problems raised by the preparation include
the necessity to be able to measure and to manipulate
picoliter volumes of liquid; then to dry them in such
a way that we get very small crystals from the content, about 1 μ, homogeneously dispersed on a dry spot;
and lastly, to end up with dried spots of the same
shape and dimensions for a given analysis.

METHOD

The samples, covered with paraffin oil to avoid
evaporation, are manipulated under stereomicroscopic
control. They are aspirated in a calibrated micropipette by pneumatic depression with an ordinary syringe. The micropipette is pulled from capillary glass,
producing an outside tip of about 3 to 8 μ; a narrowing is made above the tip with a de Fonbrune micro-
forge to an approximate volume. The exact volume delivered by the micropipette is determined between the
tip and the narrowing; it is measured by counting the
radioactivity of samples derived from tritiated water
of known radioactivity. Fifteen to thirty samples
may be taken up by the same micropipette, with each

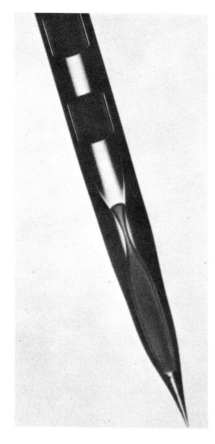

Figure 2. *Micropipette.*
The volume between the tip
and the narrowing is 170
picoliters. Three drops of
colored solution are seen
(dark) separated by paraffin
oil (clear).

sample separated from another by an oil droplet (Figure 2). By siliconizing the pipette there is no contamination of one sample by another. This can be checked by alternating in the same micropipette tritiated and normal water and later counting the radioactivity of each droplet.

The liquid samples are delivered from the micropipette onto a pure beryllium piece that has been wetted with paraffin oil. The beryllium support is engraved, to localize the samples, with 2 mm squares (Figure 3). Beryllium was chosen because of its low background properties and good electrical and thermal conductivity. Up to one thousand samples can be

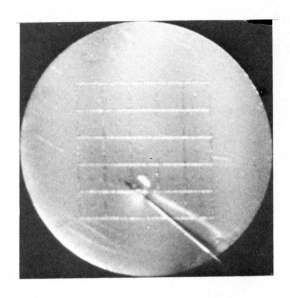

Figure 3. *Surface of a beryllium piece used to deposit the liquid droplets. It is engraved with a 1 cm square divided in 2 mm squares in order to ease the localization of the samples in the electron probe. A micropipette is seen in lower right. Bright spots are reflections on the oil covering the surfaces.*

deposited on the same beryllium piece (Figure 4).

Now follow the important steps of the preparation. We must wash the oil with a liquid of high vapor pressure, non-miscible with water, without removing the droplets. It has been found that by gently washing the beryllium block with m-Xylene for 20 to 30 seconds the oil is removed while the droplets remain on the beryllium. The beryllium piece with the liquid samples still covered by m-Xylene is then dipped immediately into isopentane which has been precooled to -160°C. This process instantaneously freezes the droplets in a quasi-amorphous state.

The samples are freeze dried under vacuum at -70°C. and below 10^{-5} Torr. Then the beryllium block

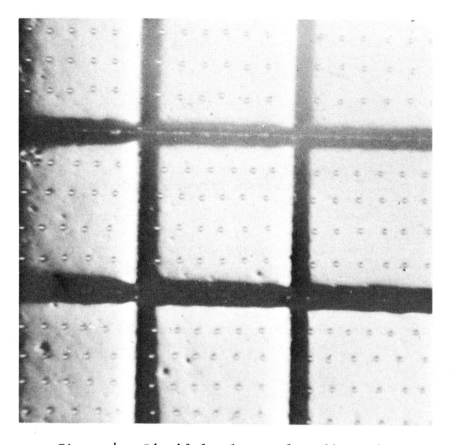

Figure 4. *Liquid droplets under oil on the surface of the beryllium. On each of the 2 mm squares
there are 24 droplets, each of 170 picoliters.*

is rewarmed under vacuum in order to avoid water condensation when the system returns to atmospheric pressure. The dried samples are then ready for the electron probe microanalysis.

The analysis is performed with a Cameca model
MS/46 microprobe. The electron accelerating potential
is 11 KeV. The beam current is fixed to 300 nA on the
beryllium and the probe is kept static. The beam diameter is enlarged to 50 to 100 μ in order to excite
the entire spot under the probe and is kept constant

during one analysis. Each sample is brought under
the beam by moving the mechanical stage under microscopic control and is then counted for 10 to 30 seconds. Four elements can be analyzed simultaneously.

Na and Mg are analyzed with a KAP crystal; P, Cl,
K and Ca with a PET or a 10$\bar{1}$1 crystal. The background
is measured on the beryllium support, off samples.

The standard solutions we use for the usual biological ions are prepared with appropriate mixed solutions of NaCl, KH_2PO_4, $CaCl_2$ and $MgSO_4$. 7 H_2O. They
provide standard curves from which we derive the concentration of unknown samples prepared and measured
at the same time under the same conditions. No corrections for fluorescence or absorption are necessary.

In order to use the beryllium again it is lightly
polished with 3 μ diamond paste and ultrasonically
cleaned in trichlorethylene, alcohol and acetone.

RESULTS

The accuracy of delivery by the micropipettes is
illustrated in Table I. Fifteen aliquots of tritiated
water were taken with 20.3 pL. and 17.8 pL. micropipettes. The radioactivity of each sample was counted
at more than 1000 counts above the background. The
absolute deviation is less than 4% of the mean. The
accuracy of delivery by the micropipettes is better
than 1% of S.E.

The uniformity of the dry spots is shown in Figure 5. The six dry spots are 40 μ in diameter. They
were obtained from 31.1 pL of a standard solution of
mixed salts, containing 100 mM Na, 105 mM Cl, 2.5 mM
of K, Ca, Mg. We see that very small salt crystals
are obtained; they are localized uniformly in each
spot on the surface of the beryllium piece; and each
spot is similar to the other one.

TABLE I

Reproducibility ^{3}H Water

20.3 pL ip/10 min	17.8 pL ip/10 min			
751	667		PICOLITERS	
728	677			
728	674		20.3	17.8
714	650			
738	619			
764	665	N	15	15
736	664			
738	640	MEAN	734	660
719	657			
735	670	SD	13.8	15.5
745	662			
711	649	SE	3.56	3.39
726	661			
743	678	% SE	0.485	0.604
731	665			

Accuracy of delivery by the micropipettes measured by counting the radioactivity of tritiated water (see text) ip⁄10 min: counts per 10
min.

TABLE II

Effect of the Rotation of the Sample

mM/L		0	90°	180°	270°	360°
		(Counts/10sec ± SD)				
Mg	10	335±19	346±23	347±16	346±19	329±12
Ca	10	263±20	257±12	260±18	260±20	267±11

11KV 500 nA 80 μ 77 pL

See text.

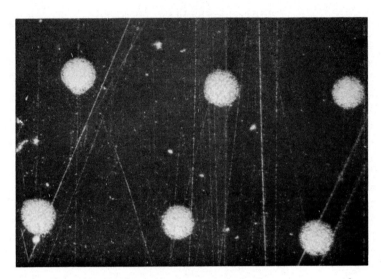

Figure 5. *Photographs of dried spots on the beryllium support. Each spot is 40 μ in diameter and was provided by 31.1 pL of a standard solution (see text).*

The homogeneity of the distribution of salt crystals in one spot is shown in Figure 6. It is an x-ray scanning picture of one 70 pL sample of mixed salts for Na, Cl, P and Mg. The scanned area for each element is 100 μ^2. We can see the absence of crystal segregation.

Table II shows that there is no significant difference in count rate for the same sample no matter what its position under the probe is in respect to the spectrometers. The sample was counted after a rotation of 0, 90°, 180°, 270° and 360° without significant differences. It means that the sample is seen as flat by the spectrometers.

Such preparation provides linear standard curves measured on five standard solutions of different concentrations of mixed salts. Figure 7 shows a standard

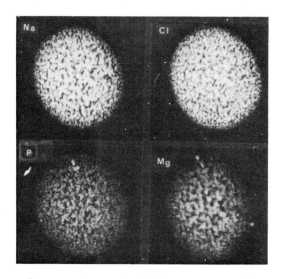

Figure 6. *X-ray scannings of one dried spot (see text).*

curve obtained from a 17.8 pL micropipette, using P concentrations of 10, 5, 2.5, 1 and 0.5 mM/L. Each point is the mean of six aliquots. The correlation coefficient of linear regression is very nearly 1 and the line nearly intercepts the origin. Figure 8 shows standard curves obtained from 20.3 pL samples of standard solution. Each sample was analyzed for the six elements, Ca, Cl, Mg, K in concentrations of 10, 5, 2.5, 1 and 0.5 mM/L; Na in concentrations of 200, 140, 100, 40 and 20 mM/L; and Cl in concentrations of 220, 150, 42 and 21 mM/L. There is a very good linearity for each of the 6 elements. Such standard curves are used to make quantitative analysis by comparing with samples obtained from the same volume of biological fluids, prepared and measured with the same method.

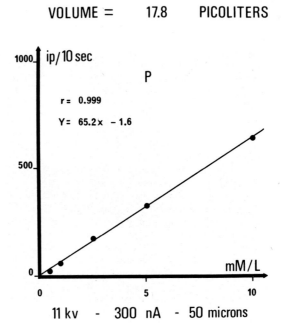

VOLUME = 17.8 PICOLITERS

11 kv - 300 nA - 50 microns

Figure 7. *Standard curve for P obtained by electron probe microanalysis of 17.8 pL of standard solutions.*

TABLE III

Reproducibility

mM/L		Counts/10sec ± SD			M
			m		
Na	200	1773±39	1723±36	1744±55 ...	1777±78
Ca	10	477±26	480±28	462±32 ...	483±27
Mg	10	216±16	210±16	208±14 ...	216±23
P	10	173±11	170± 9	166±11 ...	169±12

See text

Microsamples of Biological Fluids

VOLUME = 20.3 PICOLITERS

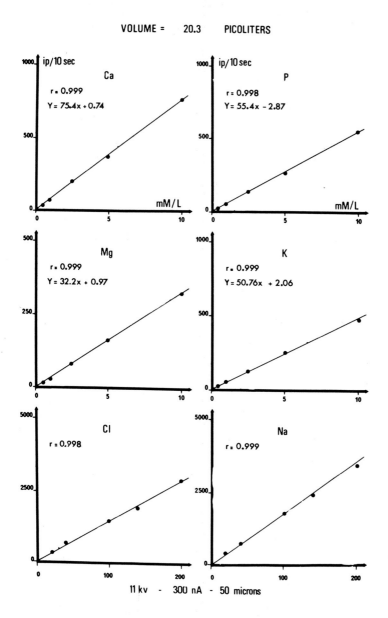

Figure 8. *Standard curves obtained by electron probe microanalysis of one 20.3 pL set of five standard solutions of mixed salts.*

C. Lechene

DISCUSSION

The reproducibility of the method has been tested
by measuring ten aliquots of the same standard solution (Table III). Each aliquot has been counted ten
times at ten seconds each for Na, Ca, Mg and P; the
mean values of such counts of three aliquots are given
in the column labeled m; then, we have taken the mean
of one of the ten-second counts for the ten aliquots;
these are given in the column labeled M. There is no
significant difference between the mean of ten ten-
second counts on each aliquot and the mean of one of
the ten-second counts on the ten aliquots. It means
that in these conditions the experimental fluctuation
is within the statistical fluctuation of counting.

TABLE IV

Effect of the Beam Bombardment

	conc mM/L	counts (ip/10sec)	duration under the Beam (min)	counts (ip/10sec)	t
Mg	2.5	313 ± 3.9	8	321 ± 4.5	−1.34 NS
P	2.5	354 ± 3.4	10	357 ± 4.5	−0.48 NS
Ca	2.5	521 ± 4.6	10	532 ± 6.2	−1.42 NS
K	2.5	568 ± 6.3	10	566 ± 6.3	0.28 NS
Na	100	2729 ± 14	7	2753 ± 14	−1.28 NS

Volume 36.6 pL   (11kv - 300nA - 50 μ)

The non-destructive characteristic of the method
is illustrated in Table IV. It shows that under these
conditions the electron beam bombardment does not affect the samples. Each element was counted for three
minutes, maintained under the probe eight minutes for
Mg, ten minutes for P, Ca, K and seven minutes for Na;
then they have been recounted for three minutes. There
are no differences between the two sets of measurements. It means that in these extreme situations the
samples are not affected during the electron beam

excitation. There is no loss of material which is a
very important advantage of the method. However, it
must be reported that chlorine can exhibit a decay of
counting rate if the beam current is higher than 300
nA for a 100 μ beam diameter. But by using a lower
current density or a higher accelerating voltage there
is no chlorine volatilization. The amount of salt
crystals excited by the electron beam does not affect
the measurement. There is no significant difference
in the intensity for P, Ca, K (0.5 mM/L) measured
either alone or in the presence of sodium chloride
(200 mM/L). It means that, as could be expected, the
contribution to the continuum brought by the amount
of NaCl (200 mM/L or 2.23 10^{-10}g/20pL) spread on the
surface of 50 μ diameter is negligible compared to
the background due to the Beryllium support.

The sensitivity of the method is shown in Table V.

TABLE V

Sensitivity

	conc mM/L	Signal (ip/10 sec)	Background (ip/10 sec)	SB
P	0.5	59.7	57.6	1.04
Mg	0.5	27.6	26.6	1.04
Ca	0.5	39.6	36.6	1.08
K	0.5	33.1	29.3	1.13
Cl	20	375	22.8	15.7
Na	20		14.1	29.5

Volume 20.3 pL (11kv - 300 nA - 50 μ)

We see that for P, Ca, Mg, and K, measured in the same
sample, the signal over background ratio is at least
one for concentrations which were 0.5 mM/L. That is
a concentration of roughly 20ppm in the liquid samples
(or 2.10^{-17}g/μ^3). It means that with the electron
probe that are presently available, the method

described easily permits one to accurately measure concentration as low as 0.5 mM/L in the original volume of material as tiny as 20 pL. There is even the possibility that by improving the design of the electron probe for biology and by preparing the samples on ultra thin support the sensitivity could be enhanced by an order of magnitude of one or two.

In conclusion, the method described provides the biologist for the first time with the possibility of quantitating the elemental contents of picoliters of fluid. The unique features of the method are the great sensitivity, the extremely small volume of sample needed, the number of elements which can be analyzed and correlated within the same sample and the relative facility of preparation and analysis. This method can be used wherever minute amounts of extracellular or intercellular fluid need to be analyzed as in renal and cell physiology, reproductive biology, arthritis, etc. The only condition being that the biological samples provide the same shape and distribution of salt crystals as the standard solutions. The same method could be used on isolated cells after mineralization. Moreover, it could be extended to organic analyses when the product of a reaction can be related to a chemical element.

REFERENCES

Castaing, R., 1951, Thesis, University of Paris, publ. O.N.E.R.A. No. 55.

Cosslett, V. E., 1966, *Modern Microscopy*, Cornell University Press, 150-156.

Galle, P., 1965, *Actual. Nephrol. Hopital Necker*, Paris, 193-206.

Gersh, I., and Stieglitz, J., 1934, *Anat. Record*, 58, 349-367.

Hall, T. A., Hale, A. F., and Switsur, V. R., 1966, In *The Electron Microprobe,* (McKinley, T. D., Heinrich, K. F. J., and Wittry, D. B., eds.), Wiley, New York, 805-833.

Ingram, M. J., and Hogben, C. A. M., 1967, *Anal. Biochem.,* 18, 54-57.

Lechene, C., 1970, *Proc. Fifth Nat. Conf. on Electr. Pr. Anal.,* 32A-32C NY, NY.

Morel, F., and Roinel, N. J., 1969, *Chim. Phys.,* 66, 1084-1091.

Tousimis, A. J., 1962, *X-ray optics and X-ray Microanalysis,* 539-557.

ELECTRON PROBE ANALYSIS OF
HUMAN THYROID TISSUE

P. S. ONG, W. O. RUSSELL, M. G. MANDAVIA, G. SROKA
The University of Texas System Cancer Center,
M. D. Anderson Hospital and Tumor Institute,
Texas Medical Center, Houston, Texas.

INTRODUCTION

The purpose of these continuing investigations is to evaluate the contribution the microprobe can make in diagnostic medicine, with emphasis on human neoplasia. The question to be answered is whether the microprobe is capable of differentiating between normal and abnormal conditions through the study of the elemental composition of fresh tissue specimens. The instrument appears to have a twofold potential:

a) by providing a sensitive, objective and discriminating source of unique data; and

b) by supplementing conventional histopathologic diagnostic methods.

MATERIALS

For this phase of the investigations, thyroid tissues were selected. Results obtained from these samples will be used as a model for further refinement and development to apply the analytical system to other tissues. The selection of thyroid tissue was not accidental. Previous investigations (Banfield *et al.*, 1971; Robison *et al.*, 1971; Robison and Van Middlesworth, 1971) have shown that elements such as calcium, phosphorus, sulphur, and iodine could be measured relatively easily. These are essential elements, and calcium and iodine content are known to be related to various metabolic processes. Tissue specimens are readily available in the hospital from patients with a variety of thyroid disorders, routinely diagnosed by experienced staff members. Furthermore, a valuable background of experience and knowledge has been accumulated from studies with other techniques, such as autoradiography. This prior knowledge of abnormal and the ill defined areas is necessary to establish whether or not the microprobe can achieve the same, or better differentiation, through determination of elemental composition of the tissues.

METHOD

Inasmuch as the objective of this project was a comparative study of the results obtained by conventional light microscopic morphologic study and the microanalytical system, as a first approach a sample preparation technique was used which would be applicable to both techniques. However, since the light microscope samples needed to be stained and embedded, separate sections from the sample were used for the two instruments.

Figure 1 shows schematically the sample preparation technique used for this study. Basically, it is the standard histopathologic method to the point where the serial sections are cut. Alternate sections were collected on silicon discs and glass slides and deparaffinized with xylene. The drying process after alcohol dehydration which is required for samples to be analyzed with the electron probe, presented some problems. Therefore, preliminary studies were conducted to compare air drying and critical-point drying techniques.

Figure 2 shows samples dried with the two techniques. As can be seen, the follicles, which consist

370

Human Thyroid Tissue

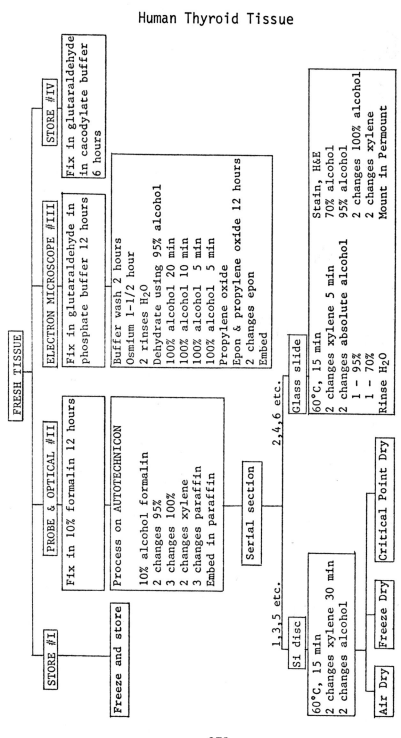

Figure 1. *Outline for tissue preparation procedure.*

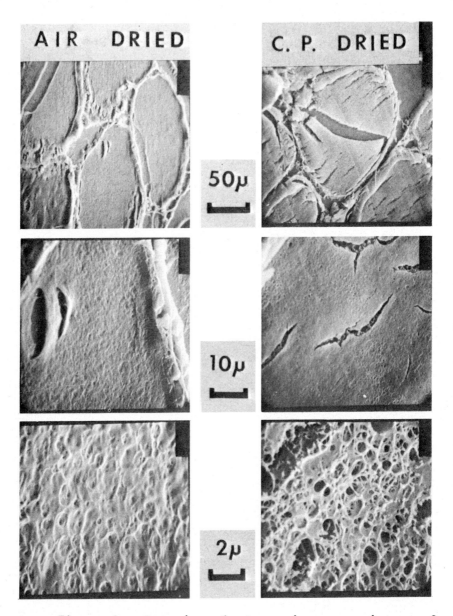

Figure 2. *Scanning electron microscope images of air-dried and critical-point dried sections of thyroid tissue specimens prepared to evaluate preservation of morphological characteristics and structure.*

basically of water, collapse when air-dried and the surrounding cells are distorted. The critical-point dried sample tends to maintain more of its overall structure although alterations such as cracks are observed, and the colloid reveals a foamy structure with a nonuniform distribution of mass, which is undesirable. Further evaluation of sample drying methods, including the freeze drying technique, is still in progress.

Figure 3 is an example of x-ray images of normal thyroid tissue and the associated optical image. Some slight distortion is obvious, although the various areas can still easily be identified.

It has been reported repeatedly in the literature that some loss of sample occurs under the action of the electron beam. In the present study, changes in intensity of x-ray emissions were observed as a function of time, and are illustrated in Figures 4 and 4a.

When a typical air-dried sample was subjected to a stationary beam, a small but rapid rise in Ca count occurred, followed by a moderately fast decrease which eventually levelled off. The corresponding silicon count from the substrate showed first an increase and then a decrease. Inasmuch as intensity of x-ray emissions from the silicon disc is a sensitive measure of the amount of material resting on it, the conclusion was that sample loss occurred, followed by an increase. The first change probably was due to water loss (at times a boiling phenomena could be observed). This was followed by a gradual moving of the protein to the periphery, which results from protein softening and electric charge. The subsequent decrease in silicon count is due to carbon contamination.

An unfocused beam does not result in changes in intensity of x-ray emissions. Shown in Figure 5 are intensities emitted by iodine and silicon. After being focused on a specific point, the beam was

Electron Image

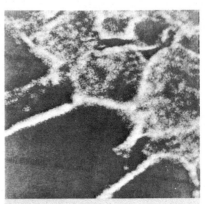

Phosphorous X-ray Image

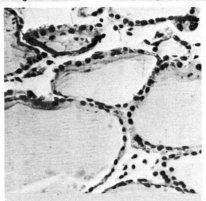

Optical Photomicrograph
of 5μ Serial Section

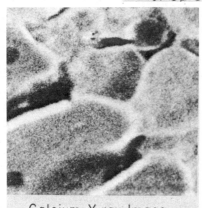

Calcium X-ray Image

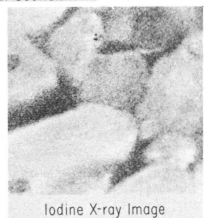

Iodine X-ray Image

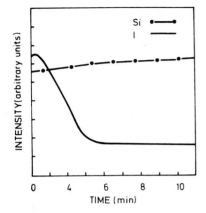 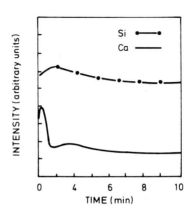

Figure 4. *A plot of simul-* Figure 4a. *Similar plot-*
taneous counts of x-ray in- *ting of counting by iden-*
tensities of iodine in a *tical method for calcium*
section of thyroid tissue *and silicon.*
and silicon in the sub-
strate measured at one
point over a period of 10
minutes.

defocused. Almost no change in silicon count was observed which reflects the stability of the sample
under the unfocused beam. Because movement of the
protein is very dependent on the softening point of
the protein and on the water content, further removal
of water was desirable. Samples subjected to a temperature of +120°C for 3 hours remained remarkably
stable under the beam even with currents as high as
1 µA.

Figure 3. *Correlation of the electron scan image*
and the x-ray images (15 µm section) of phosphorus,
calcium and iodine distribution with an optical photo-
micrograph (5 µm section) of the same area of thyroid
tissue. The apparent increased concentration of io-
dine within the epithelium of the follicle could be
caused by folds and irregularities in the section at
these areas (275-micrometer field).

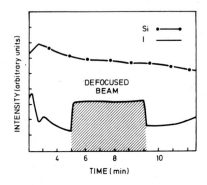

Figure 5. *A plot of simultaneous counts of x-ray intensities of iodine in normal thyroid tissue and silicon in substrate, comparing results with focused and unfocused beams.*

For diagnostic purposes, some method of quantitation is necessary. Problems arise as section thicknesses vary from day to day — even from section to section. Some tissue areas may have an open structure, and the water content can vary. Elemental measurements, therefore, had to be related to the dry mass of the tissue and complete dehydration is essential.

The method of quantitation described by Hall and Werba (1969) and Hall (1971), which uses the Bremsstrahlung of the sample for mass determination, was not suitable for these studies in which rather thick samples were used, supported on a solid substrate. (The thick substrate is necessary to serve as a heat sink). Warner's program "BASIC" (Warner, 1971) could have been applicable but required access to a computer. For practical purposes, calibration curves should be available, which an operator could use to determine sample mass. This possibility was explored by using the silicon intensity of the substrate to determine the mass of the sample.

From a particular sample, 4, 8 and 16 micrometer sections were cut. These were mounted on silicon discs and routinely processed. A specific follicle was selected in the three sections and the intensity of x-ray emission was determined for iodine and silicon. Assuming that within the follicle, the iodine concentration is constant, then a variation in intensity of

emissions for iodine in a particular section is due
to a variation in local mass, Figure 6, a plot of
iodine versus silicon intensities, shows that the
points fit on a slightly curved line which corresponds
to a certain unknown iodine concentration. A different concentration would result in a line with a different slope. Based on this method, a standard curve
was prepared as follows:

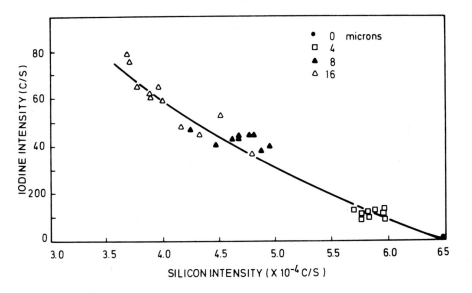

Figure 6. *A plot of x-ray intensities of iodine
versus the x-ray intensities of silicon at various
points in 4, 8 and 16-μm thick sections of normal
thyroid tissue.*

A homogenized sample of known composition was
spread at random thickness on a silicon disc and the
intensity of x-ray emissions for a particular element
plotted against the silicon intensity. Each concentration yielded one line for an element. Extrapolation and interpolation will yield a family of curves.
This method is under investigation.

RESULTS AND DISCUSSION

Three examples, of the many samples studied, are presented to illustrate the potential capabilities of this analytical system.

Figure 7 shows a series of electron scan and x-ray images of follicular adenoma of the thyroid gland. Compared with normal tissue (Figure 2), the iodine content of certain follicles is not detectable here. The phosphorus image provides a remarkably clear outline of the epithelium.

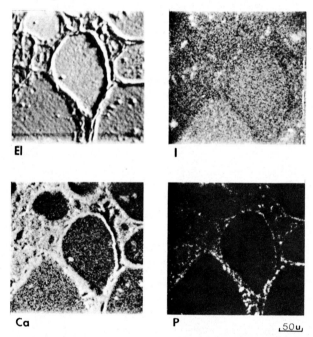

Figure 7. *Electron scan image (EI) and the x-ray images of iodine (I), calcium (Ca), and phosphorus (P) in follicular adenoma of the thyroid gland. The distribution of iodine is uniform between the epithelium and the colloid similar to that observed in the normal thyroid tissue. The intensity appears to be less than in normal thyroid tissue, but this can only be determined by the quantitative method.*

378

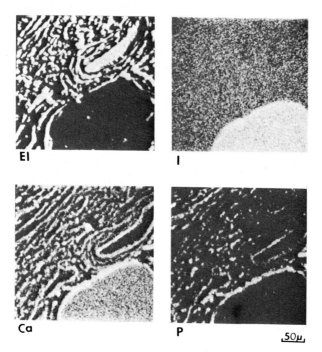

Figure 8. *Electron scan image (EI) and x-ray maps of iodine (I), calcium (Ca), and phosphorus (P) in a section of thyroid gland with papillary carcinoma. The field shows papillary carcinoma adjacent to the normal follicle. Iodine was not detectable in the malignant tissue. The epithelium and the colloid of the normal follicle stand out sharply. More iodine appears to be observed in epithelium than in colloid but quantitation is required to rule out any artifacts due to folding of the specimen.*

Figure 8 shows a series of images of papillary carcinoma of the thyroid gland. Papillary carcinoma can be observed adjacent to a normal follicle. While the iodine in the follicle is obvious, none is detected above the background in the malignant tissue.

Figure 9 shows two sets of x-ray images from the different areas of the same sample. The left row of

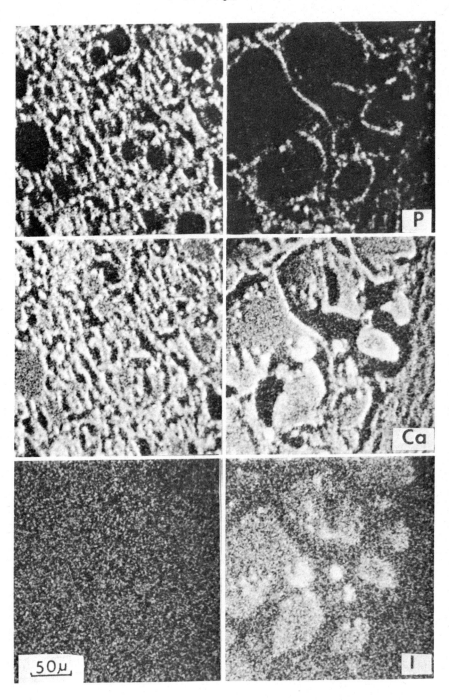

pictures represents the follicular carcinoma, the right photos are from an area which appears to be normal tissue.

While Figures 7, 8 and 9 represent two-dimensional pictures which are useful for general observation, diagnostic application requires a more precise comparison between normal, benign, and malignant tissue. Therefore, a technique of quantitative mapping of the sample was developed.

The sample was divided into a grid pattern of about 30 by 30 image points. The image points were analyzed for elemental content and the data were stored on punched paper tape. After the necessary background subtraction and quantitation, the results are presented as a matrix. Figure 10 is an example of 10 by 10 matrices for iodine.

CONCLUSION

A significant difference in the levels of iodine, calcium, and phosphorus was found in normal and malignant cells, providing new or supplementary information for diagnostic purposes. Each one of these three elements may provide the means for distinguishing between the two extremes of healthy and diseased cells, but it is clearly a correlated "profile" of these three elements combined which shows great

Figure 9. *The illustration compares an area of follicular carcinoma (left row) with an area of normal thyroid tissue (right row) from the same specimen. In the left row there is no detectable iodine, while the calcium and phosphorus ditributions outline the follicular pattern of the lesion. The x-ray images of the iodine (I), calcium (Ca), and phosphorus (P) distributions seen in the right row are comparable to previous illustrations of normal thyroid gland.*

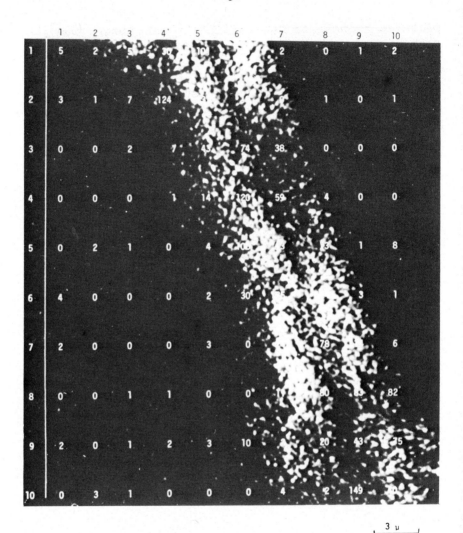

Figure 10. *Phosphorus. Quantitative matrix of the x-ray intensity map of phosphorus distribution in the area depicted. Enlarged image of the x-ray intensity map of phosphorus distribution in a 30 μm × 30 μm area of normal thyroid gland.*

promise for further improving the discriminating capabilities of the instrument.

Work is in progress to explore these possibilities.

ACKNOWLEDGEMENT

This work was supported by the Kelsey and Leary Foundation Grant 166551.

REFERENCES

Banfield, W. G., Grimley, P. M., Hammond, W. G., Taylor, C. M., de Flores, B., and Tousimis, A. J., 1971, *J. Nat. Cancer Inst.*, <u>46</u>(2), 269-273.

Hall, T. A. and Werba, P., 1969, *Proc. of 5th Intl. Congress on X-ray Optics and Microanalysis*, Springer Verlag, Berlin, Heidelberg, New York.

Hall, T. A., 1971, <u>In</u> *Physical Techniques and Biological Research*, Academic Press, New York, London, <u>1A</u>, 158.

Robison, W. L., Van Middlesworth, L., and Davis, D., 1971, *J. Clin. Endocrin. Metabl.*, <u>32</u>, 786-795.

Robison, W. L., and Van Middlesworth, L., 1971, <u>In</u> *Further Advances in Thyroid Research*, Verlag der Wiener Medizinische, Akdemie, Vienna, <u>2</u>, 891.

Warner, R., 1971, *Proc. of 6th Natl. Conference on Electron Probe Analysis*, Paper No. 42, Pittsburgh, Penn.

RECENT MICROPROBE STUDIES WITH AN
EMMA-4 ANALYTICAL MICROSCOPE

T. A. HALL, P. D. PETERS AND M. C. SCRIPPS
Cavendish Laboratory, Cambridge, England

INTRODUCTION

In this paper, after a brief description of the instrument itself, we shall describe some of the recent biological studies done with the EMMA-4 analytical microscope at the Cavendish Laboratory. We shall then consider, in the light of these studies, some of the advantages and disadvantages of the EMMA-4 in comparison with other types of electron probe instrumentation.

BASIC FEATURES OF THE INSTRUMENT

The instrument is shown schematically in Figure 1. The column is that of an AEI EM-802 electron microscope, with the addition of a special "minilens" above the specimen to provide a finely focussed beam. The electron optics of the standard EMMA are designed to provide currents up to 10 nA into a spot of diameter 130 nm (1300 Å) at a column voltage of 100 kV.

The two diffracting-crystal spectrometers (the 'ears" in Figure 1) can be operated simultaneously to analyze the generated x-rays. The original EMMA did not include a Si(Li) x-ray spectrometer, but physically this can readily be attached at the front port, and it can be used either simultaneously with the two diffracting spectrometers or separately. Several EMMAs

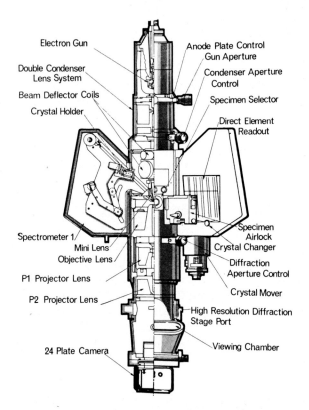

Figure 1. *Diagram of the EMMA-4 instrument.*

now have Si(Li) spectrometer attachments. The one on
the Cavendish EMMA is retractable and can be brought
to 65 mm from the specimen at closest approach; the
area of the detector chip is 80 mm^2.

The standard EMMA has no facilities for the production of scanning images. For microprobe analysis
one views the conventional EM image, selects the microarea to be analyzed, and focusses the beam onto it by
adjustment of the minilens excitation and the beam
deflector coils. An objective aperture is used if
needed to give enough contrast to see the specimen
adequately, but during x-ray analysis this aperture is
usually withdrawn to avoid excessive x-ray background.
If the specimen cannot be seen well enough to

position the probe after the aperture is withdrawn, the procedure is to align the selected microarea with the center marker on the viewing screen while the aperture is in place, and then to withdraw the aperture and focus the probe onto the center marker. The mechanical and electron-optical stability are quite good enough to permit this procedure without difficulty due to shifts.

SOME RECENT BIOLOGICAL STUDIES

One major category of EMMA studies is the tracing of material deliberately introduced into organisms. An example is the localization of gold after the administration of gold compounds used in the treatment of arthritis. Typically when one looks at micrographs of such material there is a tendency to assume that electron-dense microareas contain the injected material, but instruments like the EMMA are showing how often this assumption is unreliable. Figure 2, from a study by Yarom *et al.*, (to be published), shows electron dense areas in micrographs of sections of biopsy material taken from the knees of a patient one year after the injection of colloidal gold. The electron-dense aggregates seen in the upper micrograph were gold-positive while those in the lower micrograph were gold-negative (but osmium-positive, the block having been osmicated in fixation). Figure 3 shows other areas with their x-ray spectra. The lower spectrum, obtained with the SI(Li) spectrometer, unambiguously confirms the presence of gold (while the Cu lines were from the grid and the specimen rod, and the "Os L-alpha" line is actually indistinguishable from the Cu K-beta line). The upper spectrum was obtained with an LiF diffracting crystal, and one point of interest is the appearance of the small peak at the mercury position, arising presumably from the decay of the administered radioactive gold. (The line labeled "Os" is again not actually resolved from the copper K-beta radiation.) The EMMA study has shown on the one

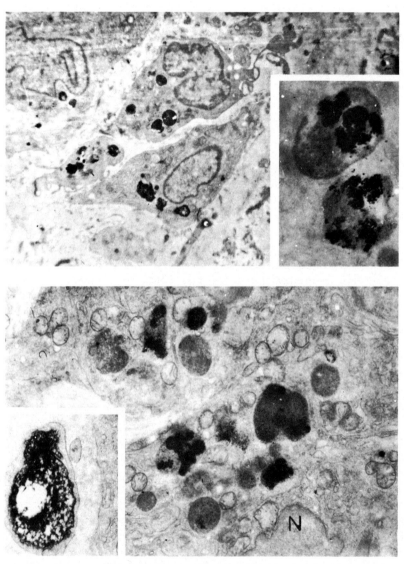

Figure 2. *Micrographs of sections of biopsy material after injection of colloidal gold, osmium fixation. Upper half: gold-positive. Lower half: gold-negative. Courtesy of Dr. R. Yarom.*

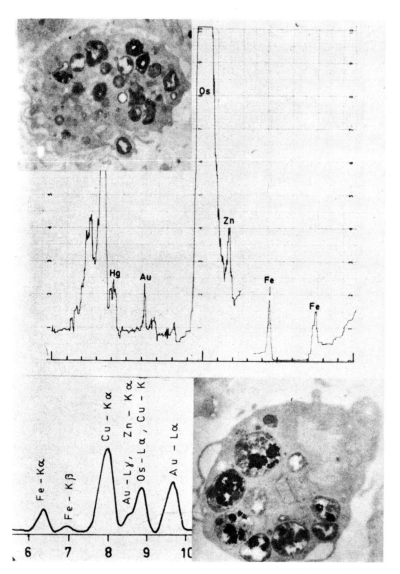

Figure 3. *Sections of biopsy material and their spectra. Lower half: Si(Li) spectrometer. Upper half: Diffracting spectrometer. Courtesy of Dr. R. Yarom.*

hand that many of the electron-dense fields do not actually contain gold; on the other hand it established that gold injected intra-articularly often lodges in the body far from the site of injection, notably in liver, spleen, and kidney, and in other bone joints.

Another example of the localization of exogenous material deliberately introduced, this time in physiology, is the study of barium bound after the exposure of tissues to barium salts (Somlyo *et al.*, to be published). In this work barium has been used as a calcium analog. Figure 4 shows a section of smooth muscle from rabbit cardiac portal vein, exposed to barium salts and stained with uranyl acetate. This section was analysed in the EMMA with the intention of confirming that the deposits in the mitochondria did indeed contain barium, but it was discovered that they

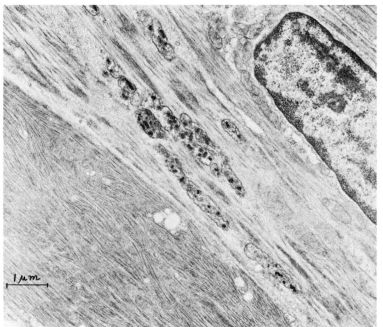

Figure 4. *Section of smooth muscle, rabbit portal vein, after exposure to barium salt and uranyl acetate staining. Barium was not detectable in the mitochondrial granules. Courtesy of Dr. A. Somlyo.*

390

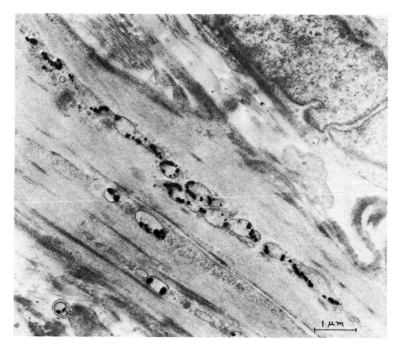

Figure 5. *Similar to Figure 4 but unstained. The granules are rich in barium. Courtesy of Dr. A. Somlyo.*

contained uranium instead. Figure 5 is of an unstain-ed section, which yielded the spectra of Figure 6. One spectrum, with the probe focussed on a group of granules, gives clear Ba peaks; the other spectrum, with the probe displaced from the granules, has no Ba peak. The study demonstrated that staining with ur-anyl acetate removed the barium and replaced it with uranium. The clarification of such aspects of stain technology will undoubtedly be a major area of the bio-logical applications of the EMMA instrument.

It is noteworthy that in the case of the barium study, the localization could not be established by means of the diffracting crystals. The available x-ray intensities were very low and analysis was prac-tical only with the higher x-ray collecting power

391

afforded by the Si(Li) spectrometer.

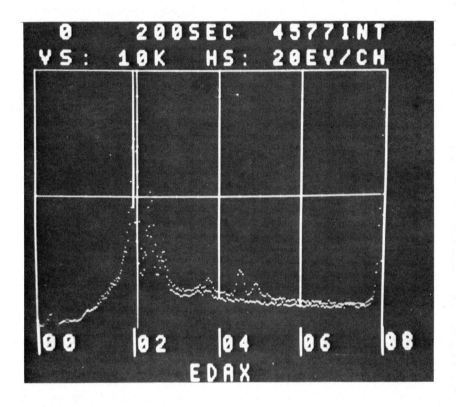

Figure 6. *Spectra from a mitochondrion of Figure 5. Upper trace: Probe focussed on granules. The two main barium L lines are clearly evident near 4.5 keV. Lower trace: Probe displaced from granules. Barium not evident.*

We have collaborated in a somewhat similar physiological study involving the localization of exogenous calcium in squid giant axon (Oschman *et al.*, to be published). In glutaraldehyde-fixed sections of squid axon, stained with lead and uranium, discrete dense deposits are seen along a line between axoplasm and Schwann cells. Similar deposits are seen in unstained sections when $CaCl_2$, 5 mM/l, is added to all of the

preparative solutions. EMMA analysis has confirmed
that these deposits are rich in calcium, and has shown
a high level of phosphorus in them as well. Further
chemical analysis of the deposits and their environment is in progress.

Another important field is the study of exogenous material which is not deliberately introduced but
is found pathologically in tissues. With Dr. K. El-
Sayed we have extensively analyzed sections of human
kidney stones (El-Sayed *et al.*, to be published), the
main intention of this work being to characterize the
layers of organic matrix within the stones, to judge
if mineral nucleation may begin with expitaxial deposition onto components of the organic matrix. Quantitative analysis was carried out for Mg, Si, P, S, K,
Ca and Fe (with the Si and S measurements probably
affected somewhat by instrumental artefacts) and several other elements were identified in the organic
matrix, including Al, Ti and V. The composition of
some microareas, in these stones taken from residents
of the Nile delta, in fact suggested a clay-like phase
with the presence of Al, Mg and Si. Ti appeared several times, and V once. In that case, in the spectrum
obtained with the Si(Li) detector, the V K-beta line
was too weak to be detected and the V K-alpha line
was indistinguishable from Ti K-beta, but the presence of vanadium was confirmed by the spectrum from
the diffracting spectrometer (Figure 7).

I would like to describe a completely different
type of study involving the use of relative measurements of mass fractions in physiology. With R. Yarom,
we have been studying the distribution of calcium in
muscle cells in different states of activation (Yarom
et al., to be published). The heart rate of anaes-
thetized open-chested dogs was regulated by a pacemaker after electrocoagulation of the A-V node, to
give normal, very slow, and very fast rates. Small
biopsies were taken from the left ventricular myocardium after ten minutes at each rate, and fixed by

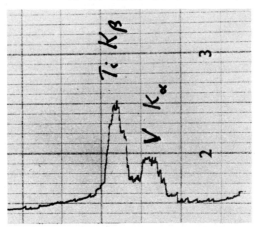

Figure 7. *Titanium and vanadium lines in the spectrum from an area of organic matrix in a section of kidney stone, observed with a diffracting spectrometer. Courtesy of Dr. K. El-Sayed.*

pyroantimonate to precipitate intracellular calcium. The tissues were fixed in a solution of potassium pyroantimonate plus osmium tetroxide, and embedded in Epon. After microtomy, relative measurements of calcium concentrations in the sections were carried out in the EMMA.

Figure 8 is a micrograph showing a typical section after microprobe analysis. Although the section is unstained, the image is adequate for positioning the beam, and the contamination spots show which micro-areas have been analyzed. (The diameter of the analyzed area is actually somewhat less than that of the surrounding ring of contamination.)

Table I presents the measurements of the relative calcium concentrations. Each measurement is the ratio of calcium K-alpha counts to continuum counts in an analyzed area, normalized to the same ratio measured in an ultrathin standard of apatite. Most striking is the clearly established and large difference between the calcium concentrations at the "slow" and "fast" rates, especially in the region of the filaments.

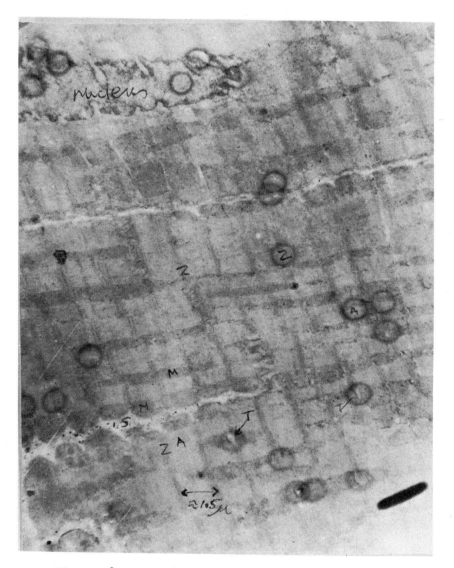

Figure 8.   *Section of unstained canine myocardium,*
after microprobe analysis.   Courtesy of Dr. R. Yarom.

TABLE I

Relative Calcium Concentrations in Selected
Spots at Slow, Normal and Fast Heart Rates.

	Slow		Normal		Fast
Thin Filaments	(12)	30±22	(6)	98±15	(17) 124±44
Thick Filaments	(9)	18±14	(5)	22±11	(13) 94±33
Triads	(13)	112±59	(5)	117±32	(10) 177±62
Mitochondria	(11)	43±12	(6)	57±16	(15) 92±47

Mean and standard deviation are given. In
the brackets are the numbers of spots analyzed.

However, biologically and histochemically the results seem to raise as many questions as they answer.
From the data one can also arrive at absolute calcium
mass fractions (using the method outlined by Hall *et
al.*, 1973) and the result is that the calcium mass
fractions in the analyzed volumes at the time of measurement were of the order of 0.1%. This is surprisingly high and it indicates that the pyroantimonate
is pulling calcium into the precipitates. There must
be a biological significance in the difference seen
in calcium concentrations after the fast and the slow
rates of activity, but the measurement of the absolute
mass fractions has shown again that histochemical techniques may do strange things.

Another recent quantitative histochemical study
is a botanical one by van Steveninck *et al.*, (to be
published), with the dual intention of determining
transport routes of chlorine and testing the reliability of Komnick's silver-staining procedure.

The plan of the study was to incubate the material (the giant algal cell *Nitella* or the water lily
in media enriched in chlorine or bromine for various
timed periods, then to precipitate the Br or Cl with
silver, and then to examine the sections. The location of the precipitates should indicate the routes

of transport. But it was important to determine the specificity of the silver precipitates: Were they all associated with Cl or Br, or might the silver be precipitating non-specifically? In other words, are the precipitates a reliable indicator of the presence of Br or Cl?

(Initially the sections were embedded in Araldite, and bromine was used in the enriched medium as a chlorine analog because the observation of chlorine itself would have been meaningless due to the chlorine content of Araldite. Even a commercial "chlorine-free" Araldite was found to contain substantial chlorine. Later, the sections were embedded in glutaraldehyde, according to the technique of Pease and Peterson (1972), so that chlorine itself could be used in the enriched media. In fact, with embedment in glutaraldehyde and bromine enrichment of the medium, it also proved possible to observe simultaneously the presence of prior chlorine endogenous in the tissue, and the distribution of the bromine taken up from the medium.)

We shall not discuss here the specific transport routes which were indicated by the precipitates (and which could be observed in any electron microscope, with no need for x-ray analytical facilities). The x-ray analyses showed that the Ag was always associated with Br or Cl. In fact, by comparison with standards of AgBr and AgCl, we could show quantitatively that the number of atoms of Ag in a probed volume was approximately equal to the sum of the numbers of Cl and Br atoms. This type of measurement is technically simple, and it is a good illustration of the contribution that x-ray microanalysis can readily make to electron-microscopic histochemistry.

We should not leave the topic of biological studies without a mention of frozen-dried ultrathin sections. In conventional preparative procedures, involving liquid baths, there is the well known danger of removal or displacement of the elements one wants

397

to study. The danger is reduced in ultrathin sections
which are cut frozen, mounted frozen on grids and
freeze-dried, with no chemical fixation or exposure
to liquids at any stage of preparation. T.C. Appleton
has prepared many such sections and analyzed them in
his own EMMA instrument in the Physiology Department
at Cambridge University (cf. Appleton, in press) and
we have had the opportunity as well to analyze some
of his frozen-dried specimens. While the ultrastructure is not seen in such sections as well as in conventional preparations, entities such as cell boundaries, nuclei, endoplasmic reticulum and secretory
granules are easily recognized. Hence important
microanalytical work can be done with frozen-dried
ultrathin sections.

ADVANTAGES AND DISADVANTAGES OF THE EMMA SYSTEM

Favorable Features

   1. Users of "conventional" electron microscopes
are provided with specimen images which are of a completely familiar kind and which are of high quality
(high spatial resolution). Such users are immediately
"at home" with the imaging side of the instrument. A
rapid entry into serious and extensive work is facilitated also by the fact that the specimen rod holds
many (six) specimen grids at a time.

   2. In scanning microprobes, a static probe is
generally positioned by reference to a fading scanning image which is generated in the preceding moments
on the display tube. There is some possibility of a
mis-registration between the static spot and the
scanning image, and one does not see the shape or size
of the probe. Positioning of the probe in the EMMA
is direct and positive; and when the probe is finally
focussed, its image is seen on the viewing screen. It
is even possible to shape the beam with the stigmators,
producing a line focus for example, so that to some

extent one may shape the probe to accomodate to the contours of the area one wants to analyze.

3. At 100 kV, which is available on the EMMA, the electron-scatter and beam spread within the specimen are much less than occur at lower beam voltages.

4. The availability of diffracting spectrometers is a big advantage in comparison with instruments which have only a Si(Li) spectrometer. We find that we continue to use extensively both types of x-ray spectroscopy, wave-length selective and energy-selective. In situations where a fairly high current can be used (i. e., where the probe need not be very fine and beam damage is not prohibitive), where attention focusses on one or two elements, and especially when the local mass-fractions are relatively low, the diffracting spectrometers actually accumulate information much faster than a Si(Li) system, which becomes limited by its excessive dead time and its poorer spatial resolution. For example, in the study of calcium in muscle, described above, a hopelessly long time would be needed to get the data of Table I from a Si(Li) spectrometer system.

Unfavorable Features

1. The unavailability of a scanning image and of a very fine probe: At present we are limited in specimen thickness to what can be viewed adequately at 100 kV in the "conventional" transmission mode. One can visualize much thicker specimens by scanning transmission even at lower beam voltage, and thicker specimens give higher x-ray intensities. We are working now to add a facility for scanning images to our EMMA.

However, we do not expect to get high-resolution scanning images, because the electron optics of the EMMA are not designed for the formation of focussed beams much under 100 nm in diameter, at least with standard electron guns. In general this does not seem

to be a serious disadvantage, because in most work
beam-spread limits the spatial resolution of the x-
ray analysis anyway to a similar value, not much under
100 nm. However there are situations where one could
do finer x-ray analysis in a very thin specimen, given
a finer probe and a very high x-ray collecting power,
and the EMMA will not be able to do this even with
the addition of beam scanning and a transmission elec-
tron-detector.

2. Limited collecting power of the Si(Li) detector: With a closest approach of 65 mm, the EMMA
Si(Li) detector has an x-ray collecting power much
less than can be had in scanning microscopes, in spite
of our relatively large detector chip (80 mm^2). Again,
in most of our work this is not a serious limitation
as we usually do not want a closer approach, and one
can get better peak-to-background ratios if the detector is away far enough to allow deflection of backscattered electrons before they hit the detector window, as is done in the EMMA with small pairs of perm-
anent magnets around the x-ray takeoff cones. But
the limited collecting power of the Si(Li) spectrometer is one of the features precluding the achieve-
ment of finer x-ray analysis in the special situation
mentioned above.

3. The EMMA beam can be focussed to somewhat
less than 100 nm, but when such a beam is spread out
to view the field, the current is too low to form a
bright image on the viewing screen. Microanalysis
with such a fine probe is therefore awkward and tiring.

4. In comparison with scanning microscopes, it
is more difficult in the EMMA to reduce or control
the extraneous continuum radiation generated near the
specimen by scattered electrons, because of the bulky
metallic objects which must be present near the specimen. This makes it more difficult to use the method
of quantitative analysis based on normalization to
the continuum radiation.

5. The limited space around the specimen position makes it relatively difficult to design facilities like a cold-stage and associated specimen transfer devices.

6. The EMMA-4 as currently constructed is not well suited for studies involving very soft x-rays (wave lengths more than approximately 1.2 nm) because there is no provision for pumping the spectrometer space to high vacuum and therefore a window is needed between the electron column and the x-ray spectrometers. This limitation is not inherent in the concept of the EMMA, and the construction should be modified to bring the spectrometer space to high vacuum.

From a technical standpoint the foregoing list of features is diverse, but from the standpoint of a range of applications a consistent pattern emerges. For the various technical reasons cited above, the EMMA is not the ideal instrument for work with low beam currents and very fine probes, or for work where beam damage must be avoided by all means available. The EMMA instrument is especially well suited to the microanalysis of features which are seen in conventional sections in conventional electron microscopes, especially features like precipitates which usually have good stability under the probe. Included as part of this range of studies is the large and important field of the evaluation of histochemical techniques.

REFERENCES

Appleton, T. C., (in press), *J. Microscopy*.

El-Sayed, K., Hall, T. A., and Newling, T. D., (To be published).

Hall, T. A., Anderson, H. C., and Appleton, T. C., *J. Microscopy*, 99, 177-182.

Oschman, J. L., Hall, T. A., Peters, P. D., and
    Wall, B. J., (to be published).

Pease, D. C., and Peterson, R. G., 1972, *J. Ultra-
    struct. Res.*, <u>41</u>, 133-159.

Somlyo, A. P., Somlyo, A. V., Hall, T. A., and
    Peters, P. D., (to be published).

Van Steveninck, R. F. M., Van Steveninck, M. E.,
    Hall, T. A., and Peters, P. D., (to be published).

Yarom, R., Hall, T. A., Peters, P. D., Rogel, S., and
    Scripps, M. C., (to be published) [on calcium
    in muscle].

Yarom, R., Hall, T. A., Peters, P. D., Scripps, M. C.,
    and Stein, C., (to be published) [on localization
    of therapeutic gold].

DISCUSSION

COMMENT: Are you going to build a cold stage for the
EMMA?

RESPONSE: We will not build one ourselves but we
have participated in a preliminary discussion with
another group which is building one, and we look forward to the time when it will be available.

COMMENT: I am excited by your measurements because I
have been interested in calcium in muscle for about
12 years, and in the microprobe for about 5, so nat-
urally I have wanted to measure calcium in muscles
with the probe. But the calcium concentrations in
most muscle cell compartments are far below the detection limit of probe methods; we must realize that
for instance in the myofibril of the living muscle
cell, the free calcium concentration is in the range
of 10^{-6} to 10^{-7} M. However, there must be some

compartments in the muscle cell, particularly the sarcoplasmic reticulum, where we can hope to analyze calcium because it seems that calcium may accumulate to a detectable level there. So I am a little bit astonished about the data in your Table. Are they really meaningful amounts?

RESPONSE: I cannot argue the point. The meaningfulness is utterly controversial. But we should note two points. First, what we were looking at is calcium — there is no doubt the calcium was present in the various sites, and I think that the relative measurements are valid. Secondly, we have compared muscles in different states of activity, and the corresponding differences in calcium levels presumably have some significance. But as to the interpretation, I will not debate it with you; I do not know.

COMMENT: The result fits the idea that calcium must enter the cell more effectively if the heart muscle is driven faster. So as an overall measurement it makes sense. But if I recall correctly that the Table listed thin and thick filaments, the measurements there do not make sense because it is impossible to detect calcium at those sites in the muscle cell.

RESPONSE: I think we have to conclude that the pyroantimonate has dragged the calcium into those sites. There is no doubt that it was there at the time of measurement.

COMMENT: What about potassium; does it not interfere?

RESPONSE: The measurements were done with a diffracting spectrometer, with which there is no significant x-ray interference from potassium. Of course the antimony lines are fairly close to calcium and they would have made trouble if we were relying on a Si(Li) detector, but the Ca line is quite cleanly distinguished with a diffractor. (diffracting spectrometer.Ed).

COMMENT: Did you check for any contamination from other cations in the precipitate, for sodium, say?

RESPONSE: We looked for sodium and did not detect it, but we do not have a good diffracting crystal for sodium so our sensitivity for that element is poor.

COMMENT: I thought Sally Page said that Glutaraldehyde didn't let pyroantimonate into muscle cells (though that was striated instead of cardiac muscle). Did you say the material was fixed in glutaraldehyde?

RESPONSE: That is my recollection; I'll have to check.

COMMENT (different Commentator): I have reacted glutaraldehyde and pyroantimonate together *in vitro* and they precipitate. You can't use the pyroantimonate in fixation solutions where glutaraldehyde is present.

RESPONSE: I have not found a reference to glutaraldehyde in my notes, and I will have to check further later. [Additional response added at a later time: A brief description of the procedure has now been inserted into the manuscript. The fixative contained pyroantimonate and osmium tetroxide, with no glutaraldehyde. In parallel observations not discussed in this paper, other specimens were fixed in glutaraldehyde alone, and in osmium tetroxide alone].

COMMENT: With respect to the calcium levels to be expected in the muscle cells, I think the values in muscle quoted earlier from the floor primarily concern the activity of the calcium ion, and the free-ion concentration is found to be maybe 10^{-6}. This has nothing to do with the total concentration of calcium, which is what the probe measures.

COMMENT: In the case of myofilaments the fact is that there is only free calcium present; there is no calcium binding to any substantial extent. I am not worried about calcium in mitochondria; I would agree if somebody found non-free calcium accumulated there. From physiological considerations one would expect to have calcium levels high enough to detect by micro-probe techniques in the sarcoplasmic reticulum sur-rounding each myofibril, particularly in skeletal muscle, and maybe in the mitochondria of the heart muscle cells.

COMMENT: I have one comment from our own work. We are using an EM 300 and in order to get down to low enough background levels, we have had to machine the entire rod at the end of the specimen holder out of beryllium. That seems to get the background down pretty well.

But I have another question; What is the best energy range to integrate the continuum over, if you are going to use the peak-to-background method, to get fairly linear continuum *vs.* mass-thickness?

RESPONSE: You want a high range, especially with ultrathin sections where you do not have to worry about the probe electrons losing much energy in the section. If you stick to the high end of the con-tinuum, the extraneous background will be minimized, because that comes from bulk material and is biased towards the low end of the spectrum. The upper end of the band might reasonably be at least 10 keV below the energy of the beam — this is safely under the energy of the beam and under the energy of the elec-trons passing through the specimen even at the low-energy end of their trajectory. This also takes you out of the region of the main characteristic lines. On the low end of the continuum band, you need only to set the level low enough to get a reasonable count rate. Unfortunately if you use a gas-flow counter and you limit the band to very high energy, you get a very low detection efficiency. With the Si(Li) detector systems another point arises. If you operate the sys-tem over a keV range which is reasonable for the char-acteristic lines that you usually want to observe, a

range which spreads the lines out nicely, you are not
collecting any spectrum in the high-energy band where
you want to count continuum. It would be very nice if
the companies would start offering an extra high-
energy channel in some way, to take care of this problem.

The ideal band, I am sure, is at pretty high
energy: For example if you are running at 60 kV, then
a quantum band of, say, 30-40 keV.

[Additional response added later: As to the other
part of the question, linear proportionality between
continuum intensity and mass-thickness, this linear
relationship is not actually required for the measurement of mass-fractions. Linearity with respect to
mass-thickness first breaks down, for both character-
istic-line and continuum intensities, for the same
reason: electron scatter, which leads to a non-linear
relationship between mass-thickness and the length of
the average electron trajectory in the specimen. But
well beyond the point of deviation from linearity
versus mass-thickness, the ratio of characteristic
intensity to continuum intensity remains independent
of mass-thickness, so that the peak-to-continuum
method is still usable].

COMMENT: Working on ultrathin sections, we still have
the major problem of sensitivity. Concerning specimen
thickness, when do you have the highest sensitivity?
Is it with sections thinner than 100 nm (1000 Å) or
thicker? Do we want thick ultrathin sections, or thin
ultrathin sections?

RESPONSE: So far as the characteristic intensity is
concerned, if we are not concerned so much with background (to which I'll return later), we think it is
best to work at high kV, 100 kV if you have it available, and then to use the thickest specimen which is
compatible with seeing what you want to see. For a
section of given thickness, at high kV you can make

up for the lowered value of the ionization cross section by using a higher probe current, without heating the specimen more because the rate of energy loss goes down; hence you can end up with the same spot size, x-ray signal and heat dissipation. But you have better beam penetration so you are free to go to a somewhat thicker section, and get a higher counting rate.

However, exogenous background is worse at higher kV. In cases where this is a limiting factor, you have to compromise at some intermediate kV. So we use high kV if we are not in trouble with background, and we come down in kV if we are.

COMMENT: And you are cutting sections approximately how thick?

RESPONSE: Usually between 100 and 200 nm.

COMMENT: There are two factors, the high voltage and the thickness of the section. At 40 kV, for example, have you best results with thick or with thin ultra-thin sections?

RESPONSE: It depends on the resolution which the problem demands. With thick sections the image may be too confusing.

COMMENT: But if the section is too thick, according to our experience, we must have a very low beam intensity; otherwise the section is destroyed. But if the section is very, very thin we can work with very high beam intensity and even if the volume in the beam is much smaller, the resultant intensity seems to be better.

RESPONSE: Well, our experience has been that we do not seem to suffer much from damage and we tend to use a section as thick as we can, so long as we can see our way around in it. We feel that we get more

signal in this way. But we have not made a definitive
study of this point.

ANALYTICAL METHODS APPLIED TO
SEM STUDIES IN DERMATOLOGY

J. P. CARTEAUD*, P. R. GRANT** AND M. D. MUIR**
*Centre Anti-Venerien d'Argenteuil,
Argenteuil, France
**Department of Geology, Royal School of Mines,
Prince Consort Road, London, England

INTRODUCTION

Of the various analytical methods recently introduced into scanning electron microscopy, we will here
discuss x-ray microanalysis and cathodoluminescence
(CL) because they have already been applied to the
study of dermatology.

In common with other collectable signals emitted
by specimens under electron beam bombardment, these
two offer considerable advantages, despite some loss
of quantitative accuracy, such as the fact that they
are relatively non-destructive and can be used to correlate morphological spatial relationships with elemental distribution.

X-ray microanalysis, which is finding more and
more extensive application in biological research,
has already been suggested as a promising tool in
dermatology by Duprez and Florentin (1970) who gave as
an example the determination of silver in a case of
Argyria. They quoted the possibility of artificially
introducing selected heavy metals as a new technique
in histochemistry. However, this publication still
stands alone in the field of dermatology.

More recently, Brown *et al.*, (1971) using x-ray microanalysis, were able to demonstrate sulphur deficiencies in a number of congenitally abnormal hair shafts. This publication too, appears to be the only one so far in its field.

This surprisingly small amount of published data has encouraged us to present our preliminary results particularly on hair, and we will try to map out the areas of future research and possible applications in the same way as one of us did for scanning electron microscopy in dermatology some years ago (Carteaud, 1970).

MATERIAL AND METHODS

Skin. A paraffin section of a skin biopsy from a case of cutaneous Argyria was examined directly without further preparation or coating.

Hair. Hairs were taken from normal Caucasian, Negro, and Asiatic subjects, and from one case of Alopoecia areata. In order to prevent removal or migration of the elements of interest, the hairs were not fixed, but merely air-dried. They were then mounted on either an aluminum or plastic specimen holder, being attached as distantly as possible from the point of analysis, on each side of the specimen holder. The axis of the hair shaft was tilted to 40° in the specimen chamber for the x-ray microanalysis.

Artificial introduction of heavy elements into hair was accomplished by two methods:

a) arsenic was taken internally (10 drops of Fowler's solution) and the hair specimens collected before ingestion (for the control sample), two hours after ingestion, and 24 hours after ingestion. The hair was then mounted and examined as described above.

b) two anti-dandruff shampoos were applied to normal hair according to the shampoo manufacturer's instructions. These shampoos contained zinc (in the form of an organic complex — zinc pyrithione) and selenium sulphide. The hairs were collected wet, both while the shampoo was on the hair, and also after thorough rinsing of the hair. The root and tip of the hair were kept separate. The hairs were then air-dried and mounted and examined as before.

All the specimens were examined using the secondary emissive mode of the Cambridge Scientific Instrument Company's 'Stereoscan' Mark IIA, before and after x-ray microanalysis, in order to determine the extent of beam damage (if any).

Coating. In the initial experiments the hair samples were coated with aluminum because, as a result of specimen charging, it proved impossible to examine them uncoated. However, because the selenium L lines are very close to the aluminum K lines, it proved difficult to resolve the two. Later experiments were carried out using carbon as a coating material.

The skin sample was examined uncoated. For cathodoluminescence experiments, both coated and uncoated materials were examined.

Equipment Used. The x-ray microanalysis was carried out using an Ortec solid state Si(Li) energy dispersive x-ray microanalyzer attached to a Stereoscan Mark IIA. The output from the spectrometer is displayed on the cathode ray tube of a Northern Scientific Econ II multichannel analyzer, and can be printed out either on a pen chart recording or on punched paper tape.

In order to minimize specimen damage, the incident beam current was kept very low — of the order of 10^{-10} or 10^{-9} A at 20 kV and in order to accumulate a sufficient number of counts, a counting time of 1000 seconds live time was selected. In each case the background (specimen holder) was also analyzed.

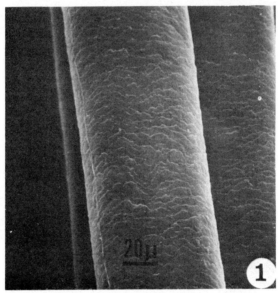

Figure 1. *Normal hair before x-ray microanalysis.*

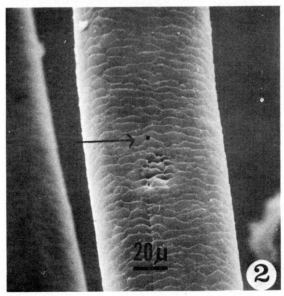

Figure 2. *Normal hair after x-ray microanalysis.*
The black spot indicated by the arrow is the point of
entry of the beam. Notice the damage below this point.

412

RESULTS

Skin. In the same way as Duprez *et al.* (1970) were able to detect silver in the Argyria liver biopsy, we were able to detect silver in our analyses of the skin biopsy (Hodges and Muir, this Symposium, Table IV). It proved impossible to localize the silver or to make any kind of distribution map, but in the light optical microscope, the silver can be seen to be localized in the sweat glands.

Hair. Figure 1 shows a human hair before x-ray microanalysis and Figure 2 the same hair after analysis. In the micrograph, the entry point of the electron beam is marked with a black dot and has an arrow pointing to it. The analysis was carried out for 1000 seconds under the conditions outlined above. The specimen was tilted at 40° with respect to the electron beam and an area of damage can be seen below the point of entry of the beam. The damaged area can be seen at higher magnification in Figure 3 and it can be seen that the scales on the hair surface have been displaced in an outward direction showing that damage occurred from the center of the hair outwards. It is possible to make a rough calculation geometrically of the volume of beam damage by assuming that this volume is spherical and calculating the volume of the sphere from the area of the intersection of the surface of the hair with the sphere of damage (marked on Figure 3 with a dotted line). This volume is surprisingly large and has a diameter of approximately 20 μm. The volume of beam penetration is presumably even larger than this since not all of the beam penetration effects will cause damage. Of course, this is a gross simplification of beam penetration and its effects, but it does serve to show that the amount of scattering in biological materials is large.

Substrate Effects

For the hair samples, aluminium proved to be an

413

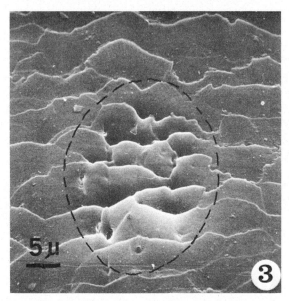

Figure 3. *Higher magnification view of the damaged area in Figure 2. The dotted ellipse represents the damaged area (from which the rough calculations were made).*

unsatisfactory substrate because of the interference of its K emission with the selenium L lines. The plastic used as a substitute was a white material that contained titanium as a pigmentation agent. This, however, was the only element detectable in the plastic and it proved suitable for our experiments.

Analyses of Normal Hair

Table I shows a comparison of the Caucasian, Negro and Asiatic hair sulphur content. The analyses of all three were carried out under identical conditions and can be regarded as comparable, although no absolute figures can be derived from the data given here. It is clear that the sulphur content of the Caucasian hair is much lower than that of the Negro and the Asiatic (Chinese) hair.

TABLE I

Relative Concentrations of Sulphur in
Caucasian, Negro, and Asiatic Hair

Sample	Peak	Back-ground	Peak–Bkgnd	$\dfrac{\text{Peak}}{\text{Bkgnd}}$	$\dfrac{\text{Peak-Bkgnd}}{\text{Background}}$
Caucasian S Kα	724	514	210	1.4	0.4
Negro S Kα	8153	1414	6739	5.76	4.76
Asiatic S Kα	23093	3036	20057	7.60	6.60

Analysis of Abnormal Hair

Analysis of the hair of a Caucasian suffering
from Alopoecia areata was carried out and although
the final figures comparing the sulphur content of
the abnormal with the normal hair are not yet available (because the experiment was carried out at a
lower count rate), the hair appeared to be sulphur
deficient.

Treated Hair

Arsenic. Although the amount of arsenic ingested
was relatively not very great compared with the
amounts used in the treatment of a variety of complaints
it was detectable in the hair sample taken 24 hours
after ingestion.

Shampoos. Table II shows the concentration of
zinc in the control shampoo, and in the hair root and
tip, before and after rinsing. Again all the analyses were carried out under identical conditions and
can be intercompared. There was no zinc detectable
in the control (untreated) hair. The zinc Kα peak/

background ratio was higher in the control shampoo than in the treated hairs. Count rates for both root and tip were consistently lower for the rinsed specimens. Concentration of zinc (as evinced by the count rate) was higher in the region of the hair tip, but so was the background. The peak/background ratio was consistently slightly higher for the hair root than for the tip.

TABLE II

Concentration of Zinc in Hair
(from anti-dandruff shampoo) at
Tip and Root, and before and after Rinsing

Sample	Peak	Background	Peak−Bkgnd	Peak/Bkgnd	(Peak−Bkgnd)/Background
Control hair			None		
Control shampoo	496	393	103	1.3	0.26
Hair tip before rinse	1078	994	84	1.08	0.08
Hair tip after rinse	811	750	61	1.08	0.08
Hair root before rinse	517	458	59	1.1	0.12
Hair root after rinse	492	393	99	1.2	0.25

Because of the difficulties of analysis of selenium on an aluminum stub, the results are not presented numerically here. However, the preliminary indications from the plastic mounted specimens suggest that the pattern of distribution is the same as it was for zinc.

DISCUSSION

Skin

A large number of elements can be deposited in the skin. Besides silver, gold, mercury, bismuth, and aluminium are often deposited as a result of cosmetic or medical treatment. Two other elements, zirconium and silicon, both granuloma-producing agents could potentially be detected and the latter of these is present in silicone plastic injections. X-ray microanalysis should be able to detect this as well as traces of the metal catalysts generally used in plastic prosthesis.

Skin biopsies, prepared by the adhesive tape method, from subjects suffering from contact allergies can be analyzed for chromium, cobalt, and nickel, which are the most usual metallic causal agents. Iodine and bromine in certain toxic eruptions, copper and manganese in psoriasis, and also the calcium content of collagen in collagen diseases are of great potential interest.

We will not discuss here the detection methods for electrolyte elements suggested by Dörge *et al.*, (1972).

Hair

Arsenic is of particular interest because it is detectable in small amounts. For instance, it can be used as a marker to estimate hair growth: X-ray distribution maps observed at the same time as secondary emissive mode views of the scales can give an

417

indication of the rate of growth. Arsenic has, of
course, an obvious forensic interest and even a significant historical importance.

The detection of thallium in hair, too, has gained a new forensic importance in view of recent reports
of poisoning by thallium.

The sulphur content of hair has, at least in our
preliminary work, a racial significance and our examination of abnormal hair bears out the results described by Brown *et al.* (1971) showing that at least
some types of hair abnormality are correlatable with
sulphur deficiency. Later work by Brown and his col-
leagues suggests that there is also a correlation between the sulphur content of hair and its parasites
(1973).

The zinc and selenium analyses of the shampooed
hair suggest that the anti-dandruff agent is actually
concentrated nearer to the scalp at the root end of
the hair. Such analysis could be of great value in
the determination of the use or abuse of such shampoos by patients in dermatology clinics.

All these applications are of considerable value
even if the method is not being applied quantitatively.

Cathodoluminescence

However, light elements up to and including oxygen cannot be detected using a solid state x-ray detector, and we have previously examined the possibilities of using cathodoluminescence for the study of
such materials (see Pease and Hayes, 1966; Carteaud
et al., 1972; Muir, Grant and Hodges, 1970; and Muir,
Grant, Hubbard and Mundell, 1971).

Because of the influence of fixation on the CL
output (Muir *et al.*, 1970; Manger and Bessis, 1970;
Muir *et al.*, 1971) we have used specimens such as

skin and hair which are particularly suitable for examination unfixed.

The skin biopsies were prepared by repeated application of adhesive tape and the hair specimens
were simply air dried.

We selected materials which emit fluorescence
under UV excitation, although CL emission may not be
strictly comparable. The hair which was examined was
invaded by a fungus, the spores and mycelium of which
were strongly luminescent (Muir *et al.*, 1971). The
keratin of the hair scales did not emit CL and the
same phenomenon was observed in skin squames invaded
by *Pityriasis versicolor* (Carteaud *et al.*, 1972) where,
although the fungal spores and hyphae were brightly
luminescent, the keratinised squames did not luminesce
at all.

The phenomenon of luminescence is a complex one
and has been discussed elsewhere (Muir and Grant, in
press). Some points with regard to the present application are worth mentioning:
        a) the presence of any transition element
quenches luminescence as it does UV fluorescence,
        b) the emission spectrum may be of value in
characterizing the chemistry of some organic compounds,
        c) the addition of dyes to hair may activate
cathodoluminescence in otherwise non-luminescent
keratin.

CONCLUSIONS

Results are relatively easy to obtain using x-ray
microanalysis and cathodoluminescence using unfixed
dermatological materials such as hair and skin.
However, it is important to take into consideration
the restrictions of the methods and to try to make
reasonable interpretations of results obtained. It is
essential that multimode examination should be carried

out, e. g. x-ray microanalysis and secondary emissive mode examination plus light optical examination, etc.

These considerations will ultimately also be extended to the various other forms of signal that can (or can potentially) be collected in a SEM such as ion analysis, Auger analysis and various forms of voltage or magnetic contrast which may in the future have a biological application.

ACKNOWLEDGEMENTS

We wish to thank Professor B. Duperrat for providing us with the specimen of Argyria, and Dr. G. M. Hodges for much helpful discussion. We are also grateful to Miss Elizabeth Morris and Mr. C. J. Markham for technical assistance with the hair samples. We also thank Mr. John Pong for allowing us to use some of his hair for analysis.

REFERENCES

Brown, A. C., Gerdes, R. J., and Johnson, J., 1971, *Proc. 4th. SEM Symposium, IITRI, Chicago,* 369-376.

Brown, A. C., Gerdes, R. J., and Haley, L., 1973, *Proc. 6th SEM Symposium, IITRI, Chicago,* 651-658.

Carteaud, J. P., 1970, *Proc. 3rd SEM Symposium, IITRI, Chicago,* 217-224.

Carteaud, J. P., Muir, M. D., and Grant, P. R., 1971, *Proc. 5th Cong. ISHAM, Paris,* 13-14.

Carteaud, J. P., Muir, M. D., and Grant, P. R., 1972, *Sabouraudia,* 10, 143-146.

Dörge, A., Gehring, K. M., Nagel, W., and Thurau, K., 1972, *Pflüge Archiv.*, 322 (Suppl.) 332.

Duprez, A., and Florentin, P., 1970, *Ann. Derm. and Syph.*, 97, 546-548.

Geisler, C., and Gottinger, E., 1971, *Wien Klin. Wochenschr.*, 83, 134-136.

Manger, W. M., and Bessis, M., 1970, *7th Congres. Intern. Microscopie Electronique, Grenoble*, 483-484.

Muir, M. D., and Grant, P. R., (1974, in press), In, (Holt, D. B., *et al.*, Eds) *Quantitative Scanning Electron Microscopy*. Academic Press, London & New York.

Muir, M. D., Grant, P. R., and Hodges, G. M., 1970, *Royal. Microscop. Soc.*, *MICRO 70 Abstracts*, 195.

Muir, M. D., Grant, P. R., Hubbard, G., and Mundell, J., 1971, *Proc. 4th SEM Symposium, IITRI, Chicago*, 401-408.

Pease, R. F. W., and Hayes, T. L., 1970, *Nature*, 210, 1049.

SUBJECT INDEX

A

Absorption, X-ray 206-, 230, 252, 254, 270
Acetone 109, 165, 358
    coolant 315
Acrolein 296
Adenoma 378
Agar-agar 214, 229
Air-drying 158, 175-, 206-, 279, 315, 370-, 410, 419
Air-lock 173, 177
Albite 156, 231
Albumin 17, 21, 124-, 130, 136, 156, 214, 229, 340, 343
Alcian blue 314
Alcohol 95, 164-, 358, 371
Algae 34-, 396
    blue-green 34-
Alkali elements 213-. See also *electrolytes*, *ions*
Allergies 417
Alloys 86
Alopoecia areata 410, 415
Aluminum 8, 10-, 20, 48, 103, 112, 222, 259, 268, 274, 283-
    coating 35, 161, 165-, 173, 212, 243, 262, 393, 411, 417
    specimen supports 243, 290, 410, 413, 416
Amiloride 347
Amoeba proteus 313-, 318-
Amphiuma 128-
Anaesthesia 393
Anions 213-, 227
Ante-chamber 173, 177
Antibody 296
Anti-freeze 160
Antimony 108, 394, 396, 403
Apatite 264, 394
Aperture, X-ray 59, 61, 67, 69, 72
Araldite 149-, 170, 233, 235, 397
Area of sampling 4, 8, 81, 269
Argon ion beams 90, 97-
Argyria 409-, 413

Arsenic 410, 415, 417
Arthritis, rheumatoid 15, 23, 366, 387
Artifacts 123-, 339, 379, 393
Atherosclerosis 15
Atomic absorption spectroscopy 21, 23, 27
Atomic number 258
Atomic number effect 254
Atomic weight 258
Auger-electron analysis 420
Autoradiography 100, 370
Axon 392
Axoplasm 392

B

Background 69, 70, 112, 145, 148-, 211, 239, 214-, 248, 285, 355, 358, 386, 405, 407, 411
Backscattered electron(s) 40, 82, 136-, 202, 258, 280, 400
Backscattered-electron scanning images 55, 111, 130, 135, 199, 269
Bacteria 34
    iron 34
    spiral 34
Baffle (cold trap) 36, 173, 193, 201
Barium 224, 390
Barley 108-
Basement membrane 277-
BASIC (computer program) 261, 264, 267, 318, 376
Beam, electron
    current 36, 246, 284, 289, 334, 342, 353, 357, 365, 385, 400-, 407, 411
    diameter 156, 247, 357, 365, 373-, 385, 394, 400-
    penetration 78-, 83, 104, 179, 223-, 248, 256, 287, 314, 413
    voltage 202, 270, 314, 342, 353, 357, 385, 399, 406-, 411
Beam, ion 3, 9, 13, 22-, 75-, 89-, 420
    penetration 81, 104
Beetles 209

Subject Index

Non-dispersive X-ray spectroscopy. See *energy-dispersive*
Nucleating agents 161, 169, 174, 204
Nucleation 204, 393
Nucleus, cell 95, 100, 138, 301-, 345, 398
Nutrient medium 314, 318-, 321, 325
Nylon (specimen support films) 161, 212, 243

O

Odontoblasts 67-
Onkolite 49, 54
Opal 51
Optical microscopy 2, 31, 35, 89, 139, 332, 371-, 413, 420
  dark field 2
  fluorescence 296
  interference 2
  phase contrast 2
Optical multichannel analyzer 26
Organelles 86, 314, 330
Organic
  mass 180, 271
  matrix 21, 63-, 77-, 207, 248, 254, 257, 393
  molecules 4
  specimens 7, 13
  standards 214-, 229-
Organometallic compounds 214
O-rings 201
Orthoclase 218-
Oscillatoria 34, 44-
Osmium 95, 145, 275
  tetroxide 109-, 125, 129, 275, 296, 310, 371, 387, 394, 404
Ossification 311
Ouabain 342-
Oxygen 200, 252, 254, 256-, 261, 266, 334
  ion beams 90
  oxides 40-, 98

P

Pacemaker (cardiac) 393
Pancreas 209
Pancreatin 278
Papillary carcinoma 379
Parabolic mirrors 39

Paraffin embedding 18, 95, 124, 279
Paramecium 34
Parasites, hair 418
Parenchyma 110-
Parylene 259, 262
Pathology 369, 415-
Peak count or intensity 158, 179-, 206-, 240, 345
  diffracting crystal 145, 244
Peak-to-background ratio 18-, 40, 77, 170, 180, 207, 243-, 282, 365, 400, 415-
Penetration of beam. See *depth of analysis*
Peptone 184, 318
Permeability 159, 209
PET (penta-erythritol) diffracting crystal 358
Phosphate(s) 224, 325, 334
  buffer 279, 371
Phospholipids 324, 327
Phosphorescence 29, 37
Phosphorus 41, 43-, 112, 178, 188-, 257, 261, 283, 317-, 343, 353, 358-, 369, 373-, 393
  phosphates 224, 279, 325, 334, 371
Phosphorylation 294
Picoliter samples 352-
Pityriasis versicolor 419
Plankton 34, 43-
Plant tissues 108-
Plasma, laser-produced 7
Plasma membrane 316
Plasmalemma 114
Plasmodesmata 115
Podocytes 133, 135
Polarization 28
Polyether complexes 213-
Polymerization 235
Polysaccharides 41, 49
Polystyrene latex 321
Potassium 10-, 63, 84, 95, 131, 136-, 145-, 155-, 158, 160, 175, 178, 182-, 206-, 213-, 229, 231-, 248, 257, 261, 275, 283-, 289, 314, 316, 319-, 337-, 358-, 393-, 403
  permanganate stain 114
Precambrian age 33, 41
Precipitation 107-, 211, 352, 394, 396, 401, 404

Subject Index

Spores, fungal 419
Spray freezing 332
Spurr's medium 109
Sputtering 85, 102
Squid 392
Stability
  electron-optical 386
  mechanical 386
  of specimens (see also *preservation of*, *damage*) 229-, 288, 364, 375
Staining 86, 277, 390, 392, 396
Standards 85, 136, 146, 156, 178-, 205-, 213-, 229-, 239, 249, 260, 266, 271-, 282, 321, 343, 358-, 377, 394, 397
  X-ray fluorescence 66-, 214
Static probe analyses 297, 305
Statistical variation 321, 364
Steel 273
Stoichiometry 227
Stopping power (for electrons) 258
Stratum granulosum 341, 348
Stroma 284
Strontium 224, 318, 321, 324
Subcellular analysis 77, 190
Sublimation, of water
  deliberate. See *dehydration, freeze-drying*
  inadvertent 79, 177
Substrates. See *supports, specimen*
Sucrose 63, 66-
Sulphides, metal 33
Sulphur 65, 112, 189, 222-, 230-, 233, 236, 274-, 283-, 343, 369, 393, 410, 414, 418
Supports, specimen 3, 94-, 178, 185, 212, 251, 259, 262-, 269, 279, 290, 373-, 413
  carbon 109, 290
  films 67, 109, 150, 156, 161, 230, 239, 242-, 251, 338
  glass 1, 287
  grids or discs 242-, 315, 355, 370, 387, 410, 413, 416
  quartz 10, 14, 130, 290
Surface tension 123
Suspensions of cells 166-
Sweat glands 413
Symplasm 115

T

Tantalum, X-ray apertures 61
Target, X-ray tube 61
Teeth, rat 67
Temperature control 193
Temperature, specimen 29, 168, 176-, 199, 203-
  effects under beam 147, 149, 151-, 153-, 158, 197, 232, 236, 407
Tendon, rat tail 246, 248
Tertiary age 56
Tetrahymena pyriformis 176, 180, 184-, 257, 313-, 317-
Thallium 224, 418
Thermal conductivity 72, 76, 157, 170, 204, 355
Thermistor 193
Thickness of section or specimen 61-, 64, 121, 148, 199, 203, 206, 237, 248, 251, 256, 261-, 267, 270-, 308, 344, 376-, 399-, 406
Thiocyanate 224
Thyroid 99-, 369-
Time differentiation, of laser-induced signal 20
Time-of-flight 11-, 14
Tin (in phosphor) 39
Tissue culture 277
Tissue sections. See *sections, tissue*
Tissue, wet. See *hydrated cells or tissues*
Titanium 309, 393, 414
  target of X-ray tube 61, 65-, 72
Tonoplast 114
Torroidal mirrors (X-ray) 73
Toxic agents 417
Trace metals 15, 81
Trajectories, electron 84, 258
Transfer of specimens 177, 192-, 401
Translocation 100, 107, 123, 175-, 201-, 297, 319, 330-, 397, 410
Transmission scanning images/microscopy 169, 195-, 269, 340-
Transport of ions 107, 115, 337-, 396
  intestinal 293-
Triad structures, in muscle cells 396
Trichlorethylene 358
Tritiated water 354, 358
Trypsin 278

Subject Index